ALICE THOMSON

The Singing Line

Born in 1967, Alice Thomson is associate editor, columnist, and interviewer for the *Daily Telegraph*. She is also restaurant critic for *The Spectator*. She was named Alice after her great-great-grandmother Alice Todd, for whom Alice Springs, Australia, is also named.

THE
SINGING LINE

Tracking the Australian Adventures
of My Intrepid Victorian Ancestors

— . — . . — . — . .

Alice Thomson

Anchor Books

A DIVISION OF RANDOM HOUSE, INC.

NEW YORK

This book is for Ed

Anchor Books and colophon are registered trademarks of Random House, Inc.

The Library of Congress has cataloged the Doubleday edition as follows:
Thomson, Alice, 1967–
The singing line : tracking the Australian adventures of my intrepid Victorian ancestors /
Alice Thomson
p.   cm.
Originally published in the United Kingdom by Random House UK.
ISBN 0-385-49059-3
Includes index.
1. Australia—Description and travel.   2. Thomson, Alice, 1967– —Family.
3. Todd, Charles Heavitree, 1826–1910.   4. Todd, Alice, b. 1836.   5. Todd family.
6. Travelers—Australia.   7. Travelers' writings.   I. Title.
DU105.2.T47   1999
994.03 21

Anchor ISBN 978-0-385-49753-4

Author photograph © Andrew Crowley

www.anchorbooks.com

Printed in the United States of America

147482399

Contents

Maps

Alice lost her virginity
Witness by
The old man gum tree
While the dog sat confused
Patiently licking its wounds
She gave birth
To one stone room
Next a shed then a house

She then stepped one step south
Before the caterpillars knew
Alice grew
With the scenery so strong
The old man gum tree
Witness Alice lose her virginity
Long before me

'Fertility'
David Mpetyane, Aboriginal artist, 1992

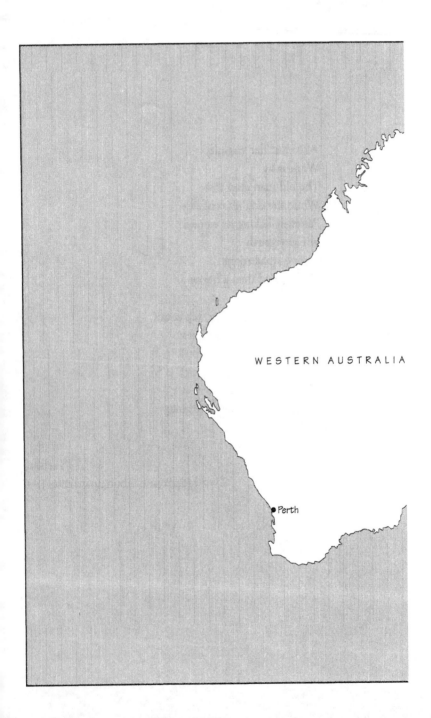

WESTERN AUSTRALIA

● Perth

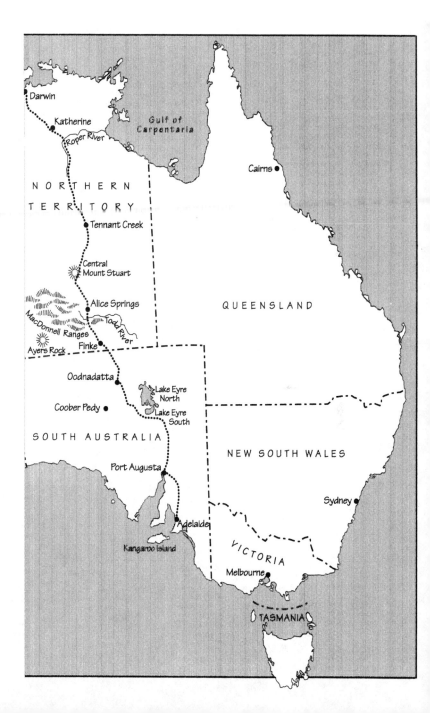

1

The Proposal

I could have been called Patience, Gwendoline, Kathleen or Maude, all family names. Instead, I was christened Alice after a solemn-looking great-great-grandmother who had black hair framing a round face, pale eyes and delicate hands. In every generation of my family someone had been named after this sepia woman, set in red velvet in our dining room. The original Alice, in her matronly Victorian crinoline, didn't look like an obvious role model. But she had one great redeeming feature; the story of her marriage proposal to a total stranger.

In 1849, when she was only twelve years old, my great-great-grandmother was reputed to have done something few women nowadays would be brave enough to consider. One of eleven children of the Bell family in Cambridge, she was alone in the schoolroom one day and bored. Looking out of the window, she saw a man twice her age with a neat beard and narrow shoulders walk up to her black and white gabled house off the marketplace in Free School Lane.

Running down to the kitchen, she was told that this skinny, pallid creature was a distant cousin who had come for 'white wine sherry' and Madeira cake with her mother. Intrigued by his forlorn face, Alice slipped into the drawing room and hid behind the chaise longue. There she listened as the awkward visitor explained that he had just been promoted to the job of assistant astronomer at the University Observatory.

The young grocer's son, a Mr Charles Todd, had been given a letter of introduction to his wealthier merchant cousins by his patron, the seventh Astronomer Royal, Sir George Airy. The formidable Mrs Bell politely inquired after the man's family, but with a depressive for a father and an invalid for a mother, Charles was unforthcoming. He and his two brothers and sisters had watched as the family's fortunes, in the form of a tea and groceries emporium in Islington, had dwindled into a four-barrel wine merchants in Greenwich.

Charles would have followed his elder brother into the merchant navy on the first ship out if it hadn't been for Sir George Airy. The famous astronomer had plucked the fifteen-year-old out of the local Roan school in Greenwich. As a governor of the school, Sir George had heard about the young boy's extraordinary talent for mathematics. By seven, Charles used to earn his pocket money in the port's alehouses, adding or subtracting lists of numbers for the amusement of customers.

The Astronomer Royal was reorganising the Observatory, and needed more human calculators to collate a mass of observations. Charles became a 'supernumerary computer' sitting on a high, backless wooden stool totting up figures with five other, better educated, boys. But he could do it far faster. Desperate to get away from his family home, the diffident youth persevered. After seven years he heard that a colleague had turned down the position of junior assistant at the University Observatory in Cambridge. Eager to study new projects rather than confirm old discoveries, Charles begged his superior to put in a good word. Within weeks he had moved to Cambridge and was spending his nights making the first observations of Eaye's Comet and searching out the newly identified Neptune.

Alice knew none of this story as she heard the twenty-two-year-old politely discussing his new position with her mother. Struggling to think of something to say, Charles explained that he had just seen the shadowy mountains on the moon through the telescope donated by the Duke of Northumberland, but had never before travelled further than London. Mrs Bell, aware that she was probably this gauche astronomer's only contact in Cambridge, asked him whether he had yet found comfortable lodgings. Charles admitted that his rooms were spartan but that he was working too hard to notice.

'You should get married, Mr Todd,' Mrs Bell suggested. 'I fear no one would want to marry such a dull fellow as I,' Charles replied. Suddenly, Alice jumped up from behind the chaise longue and, according to family legend, announced, 'I will marry you, Mr Todd, if no one else will.' There was a long silence. 'You are far too young,' said the applicant astronomer, nervously clicking his lily-white knuckles. 'You can wait for me,' said Alice.

Mrs Bell, blushing at the forwardness of her youngest daughter, sent Alice from the room. But Mr Todd was already smitten by this fleeting vision in pinafore and plaits, with her thick, black hair like a Chinese coolie, freckles and straight eyebrows. The next day, a paper package was delivered to the front door for Miss Alice Gillam Bell. Inside was a copy of the *Pilgrim's Progress*, inscribed to the young Miss Bell from her friend, Mr Charles Todd.

The book presented Mrs Bell with a dilemma. She couldn't send such an innocent gift back to the poor man. He shared the same nonconformist religious background and was an upright figure. She didn't, however, want her daughter being courted by a pauper. She was well acquainted with the precarious financial circumstances of the Todd family. Through hard work, her own husband, Edward, had risen to become one of the most successful corn merchants in Cambridgeshire. Their home at No. 3 Free School Lane had five floors, and their new warehouse in Pease Lane was the most impressive for miles. Once Mrs Bell had made bonnets to supplement the family's income; now she hoped that one day a son might become mayor of Cambridge. Mrs Bell spent occasional evenings speculating on who would be lucky enough to lead Alice up the aisle, but that would be many years off. Her adored and precocious daughter needed to become less impetuous.

To Mrs Bell's relief, Alice seemed to forget her promise for the next seven years, even though she was sent biblical tracts every birthday, with increasingly daring inscriptions signed by 'your admirer Charles Todd'. Occasionally he would call on the schoolgirl, but she was never allowed to visit his lodgings overlooking Trinity College to drink his favourite blend of Su-Chang and Orange Pekoe tea.

By the time Alice was fifteen, Charles had been sent back to Greenwich to take charge of the 'Galvanic department'. His main job was the maintenance of the time-balls which were placed at Greenwich and in the Strand in London. Charles had to ensure that these balls, suspended at the top of tall masts, were dropped at precisely one p.m. each day to provide an accurate time check. In London the balls were used by lawyers and businessmen, walking down the Strand, to check their pocket watches on the way to their clubs. But their main use was for ships in port, so that captains could set their maritime chronometers accurately for a voyage. Using the time-balls to check Greenwich Mean Time, the navigators could establish their longitude anywhere in the world by comparing their chronometer's time with an estimation of local time. An hour's difference represented a shift of fifteen degrees in longitude. The chronometer thus had to be set to exactly the right time or every minute lost or gained would mean an error of longitude of nearly twenty miles on the equator. The time-ball couldn't be a fraction of a second out.

This work brought Charles into contact with the newfangled electric telegraph. Since the Napoleonic wars, Britain had been looking for a way to communicate rapidly with her expanding empire. Throughout the 1830s, men were experimenting with magnetic needles, coils of wire and galvanic electricity. In 1838, Professor Charles Wheatstone and his colleague William Cooke patented the first long-haul telegraph instrument in Britain, where the letters were denoted by a number of motions to the left or right of the needle. In America, Samuel Morse, an artist by profession, had also formulated his ideas for transmitting messages by means of electricity through dots and dashes. In 1844, he inaugurated the first intercity line

Morse Alphabet, Numerals and Special Characters

ALPHABET

A	._	H	....	O	_ _ _	V	... _
B	_ ...	I	..	P	. _ _ .	W	. _ _
C	_ . _ .	J	. _ _ _	Q	_ _ . _	X	_ .. _
D	_ ..	K	_ . _	R	. _ .	Y	_ . _ _
E	.	L	. _ ..	S	...	Z	_ _ ..
F	.. _ .	M	_ _	T	_		
G	_ _ .	N	_ .	U	.. _		

NUMERALS

0	_ _ _ _ _	4	.... _	8	_ _ _ . .
1	. _ _ _ _	5	.....	9	_ _ _ _ .
2	.. _ _ _	6	_		
3	... _ _	7	_ _ ...		

SOME SPECIAL CHARACTERS

The bar over two or more letters indicates that they are to be transmitted as a single character.

AA	. _ . _	Unknown station call.
AR	. _ . _ .	End of transmission sign, used when no receipt is required.
BT	_ ... _	Long break. Precedes and follows the text portion of a message.
EEEEEEEE	....	Error. A succession of eight or more Es means 'Erase the portion of the message just transmitted; the corrected portion will follow': or, if followed by AR, means 'Cancel this message'.
IMI	.. _ _ ..	Repeat. Made by the recipient, to the originator. If made alone it means 'Repeat all of your last transmission'. If the sign is followed by the letters AA (all after), AB (all before), WA (word after) or WB (word before) followed by a word, then it means 'Repeat only that portion of the message so indicated'. It is used by the originator to precede the second transmission.
K	_ . _	'Invitation to transmit' or 'This is the end of my transmission to you and a response is necessary'.
R	. _ .	Received; means 'I have received your last transmission'.
AAA	. _ . _ . _	Period (in plain language, a full stop).

from Washington to Baltimore with the words: 'What hath God wrought.' In the same year, Queen Victoria used the new telegraph from Windsor to London to announce the birth of her second son, Alfred Ernest.

In Britain the telegraph was originally used as a signalling device by the train companies, but the police soon realised its potential. Pickpockets used to prey on crowds at busy railway termini and then escape by train. The telegraph allowed police to alert stations up the line of a thief's impending arrival. It was also credited with having caught the murderer John Tawell. He tried to escape from Slough by train, having killed his mistress. But the police in London were immediately telegraphed, and they arrested him when he stepped on to the platform at Paddington. Tawell was convicted and hanged, and the telegraph wires became known as: 'the cords that hung John Tawell.'

The Stock Exchange was the next to go on-line. But the public was still nervous of the wire, with some insisting that it was witchcraft. It was only during the Great Exhibition of 1851, when thirteen different telegraph instruments went on display, that people began to be excited by the idea. Morse code soon replaced Wheatstone's more unwieldy system. A wedding was conducted down the line, and there was serious debate as to whether the extension of the wires to Gretna Green would mean the end of runaway marriages, because a disapproving parent could alert the authorities before their child arrived.

In 1852, the Astronomer Royal installed a magnetic clock for the transmission of Greenwich Mean Time around Britain using the new electric telegraph. Stationmasters were issued with the order, 'You are at liberty to allow local clock and watch makers to have Greenwich Mean Time.'

The first time Charles used the telegraph was to convey the time from Greenwich to his time-balls with total precision. Charles wrote of his duties: 'The timeball is connected by wires to that of the Observatory, and when the ball falls at Greenwich, an instantaneous shock of electricity will be communicated along the wires. This, acting as an electric trigger to the ball in the Strand, will cause it to fall simultaneously with the one at Greenwich and indicate the exact time to all London.'

The grocer's son had found his great passion. Charles adored this pulsating wire that could disseminate information so rapidly across the country. He looked forward to the day when the telegraph would ensure that all the timepieces in England were ticking to Greenwich Mean Time, instead of chaffing against each other with their idiosyncratic local times.

On his days off, he would walk to the telegraph offices at the railway terminus to talk to Charles Walker, the Telegraphic Superintendent and Electrician of the Railway, and watch the messages arriving. This nineteenth century trainspotter was fascinated by the way the telegraph could control the traffic on Britain's growing railway system by allowing signal boxes to communicate. He also spent hours at the Electric Telegraph Company's office absorbing the latest developments and debating the potential of underwater cables. In 1850, the first telegraph wires, wrapped in a kind of rubber called 'gutta-percha', had been laid under the Channel between Dover and Cap Gris-Nez. The first wire broke because a Normandy fisherman caught the metal coil and thought he had discovered 'a new kind of seaweed with gold in it'. But two years later, in 1852, the operation was up and running and Britain soon made contact with Paris, Berlin and Vienna.

By 1854, over 3,000 miles of telegraph had been erected in Britain, America and on the Continent. There were lines as far afield as Cuba and Chile. It didn't take long for Charles to realise the effect that this thin metal wire could have on the expanding empire. The idea of linking continents and people with a series of dots and dashes appealed to his acute sense of order.

Alice, meanwhile, was obsessed by another code: the list of men's names her elder sisters chanted on returning from dances. There was Herodotus Hollyhead, an earnest young doctor with russet curls and a rich maiden aunt, and Endurance Smith, the pious-looking boy from the Baptist church with golden ringlets who spun her eldest sister, Sarah, round on the dance floor until she felt faint. Admirers went in and out of fashion in the Bell household. Alice, still confined to the schoolroom, would listen attentively. The name Charles Todd barely entered her head, and when it did, it was to read out to her sisters in mocking terms the endearments he sent her.

On her eighteenth birthday, when Alice received a book of hymns with the words, 'To my beloved Alice from her admirer Charles Todd', she barely glanced at it. She was too excited by the thought of her first ball. It was only after spending the night being clumsily whirled around by middle-aged family friends, and tongue-tied youths, that she became intrigued by the idea of Mr Todd. She hadn't seen him for two years and her father had mentioned that he was now gaining a reputation in the field of telegraphy.

When Mr Todd came to tea, he was quite as small and ungainly as Alice had remembered and was wearing the same black-rimmed glasses. He was far more startled by the transformation of schoolgirl into handsome woman. By now, Charles's early vision was three inches taller than he was, with smooth white hands, a steady gaze, a superior nose, and hair coiled into the nape of her slender neck. He was entranced by Alice, who hadn't yet developed the languid torpor expected of mid-Victorian women. Alice was amused by this mathematician's fine wrists and by the way that his moustache curled under his lips. Yet he seemed even more twitchy than on his first visit. Alice may not have wanted to marry him, but she wished her admirer would relax and create a good impression on her sister Sarah, who had joined them for tea.

After the Madeira cake arrived, the young man announced to Mrs Bell that he had been offered the chance of a job in the newly created colony of South Australia. The would-be emigrant became eloquent as he explained that India was swamped with second sons of gentry trying to escape the Church. Africa had too many missionaries and he had no calling. But Australia, with its gold, sheep and untouched lands, was the ideal place for a man of small means to make his name.

His new position would be Government Astronomer and Super-intendent of Telegraphs for South Australia, for which he would be paid £400 a year. Mr Todd admitted there were no telegraph lines in this new land, but he could build those, and he relished the challenge of learning about the celestial movements on the other side of the world.

Having spent the past thirteen years working so hard, he had few friends to miss him. The only problem was the question of a wife, which he had been advised to acquire before leaving, there being a shortage of eligible girls in the convict continent. But who would want to marry a mathematician, who lacked the stature of an officer, the weight of a man of God or the glamour of an explorer? Who would be prepared to go to a country thought to be populated by thieves?

'So would you be taking anyone with you?' Mrs Todd inquired anxiously. 'I cannot ask anyone to share what might be a rough and crude life,' Charles replied. For the second time there was a tense silence. Then Alice, feeling sorry for the poor man, said, to her own astonishment: 'I will go with you, Mr Todd.' For several moments no one spoke. Then Charles came to sit next to her on the chaise longue, and, having kissed her forehead, accepted her proposal.

The four months before they left were an anticlimax for Alice. She barely saw her future husband. He was busy getting all his instruments together for the long voyage and gathering information from the Electric Telegraph Company before they set sail. As she learnt more about her prospective new home, a huge, hot continent populated by convicts and 'wild blacks', Alice began to realise the significance of her decision. None of her family would have blamed her if she had changed her mind. Mrs Bell may have quietly wished that her daughter would renege, but Alice never went back on her promise.

Her mother fussed around, making the dove-grey wedding dress and a bonnet lined in pale blue satin. Her sister Sarah bought her a veil of Limerick lace. She was given the schoolroom piano, being the most musical child, the bone-handled spoons, some silver pickle jars and a mahogany sideboard. Mrs Bell also persuaded her own cherished maid, Eliza, to accompany her daughter to the new land.

The wedding was a small affair at the Baptist Chapel in St Andrew's Street on 5 April 1855. Charles made a pretty speech commending his new wife, and adding that one day he would like to see a telegraphic string stretching round the world, like the necklace of pearls around Alice's throat. The guests cheered, although they had little idea what he was talking about.

The next month, Charles told his leaving party in Greenwich that he wanted to be 'instrumental in bringing England and Australia into telegraphic communication'.

Only the explorer Charles Sturt was able to appreciate the extent of this young man's ambition when Charles wrote to him out of the blue: 'I look forward to the time when our telegraph system will be extended to join the seats of commerce in Australia.' As Sturt was well aware, the Australian continent hadn't yet been crossed.

Charles was in charge of provisions and thought of the smallest details. He bought Alice a toothbrush, a tin of boiled cinnamon sweets and a hand mirror. He even remembered hairpins for Eliza. He methodically wrote down their joint belongings and purchases in a leather notebook. Alice's father supplied them with the best tickets he could afford, two state rooms on the small ship *Irene*. With seventy tons of cargo, including Alice's piano, the barque set sail from the Downs in late June.

It anchored again in the evening off Folkestone and Charles took his new wife on deck to admire the cutters beating down the straits. This was Alice's first sight of Dover, and as there was 'a fine full moon', Charles could point out the castle and the cliffs. To Alice's delight they even glimpsed a train enter and emerge from the tunnel on its way to London.

The passage down the Channel was rough, with gales, heavy seas and rain. Alice could barely make it out of her bunk before noon. Losing sight of the Lizard, she felt unbearably queasy, but put it down to the rolling of the ship rather than any apprehension concerning her new life. By the time she celebrated her nineteenth birthday, with a cake made of the last fresh eggs and a tot of rum for the sailors, she had found her sea legs.

The story of my great-great-grandmother's proposal has been embellished by my family over the years, so that by the time it reached me there were several versions. Sometimes Alice would be lying Lolita-like on a bearskin rug, other times she would be playing Spillikins under the table. My mother knew that they were drinking 'white wine sherry', and I had added the lily-white knuckles. But in every story the words were the same: 'I will go with you, Mr Todd.'

However humourless Alice appeared in her portrait, I was happy to share her name. Other female ancestors were less admirable exemplars. One had embroidered slippers and a hotwater-bottle cover to send to Stalin and another had been the mistress of the Kaiser. Alice, I decided, had not only gambled her life in a moment of grand passion – or more probably compassion – but seven years later she was prepared to honour the bet.

It was quite by chance that I discovered that my great-great-grandmother was more than just an ancestral curiosity. My grandmother, another Alice, had bought me a jigsaw puzzle of the world for my thirteenth birthday. As we filled in the endless miles of blank space inside Australia, she spotted Alice Springs. 'This town,' she told me, 'was named after your great-great-grandmother. She was the Alice of Alice Springs.'

The young Mrs Todd had given her name to one of the most famous small towns in the world. Unlike her more worldly cousins, who had settled down with fusty academics and doctors in Cambridge, she had won herself a footnote in colonial history. Charles Todd may have looked like a spindly mathematician. He developed an appalling sense of humour, insisting every time he drank tea on punning, 'Without my Tea I would be Odd'; he often disappeared into the outback, leaving his wife and children in a settlement that began as scarcely more than a shanty town. He cared little for social etiquette and was always slightly grubby with dust from his travels or with grease from his astrological instruments. But what should have been a five-foot-five marital calamity had managed to put his wife on the map at the red heart of the new continent.

2

·· — — —

In Search of Alice

Until recently this family legend was all I knew of the story and all I wanted to know. Alice Springs, I was assured by those who claimed to know, might once have been a bonza little town but it was now a plastic paradise full of cheap thongs, garish cafés and sweaty shopping malls. When people invariably asked, 'Alice in Wonderland – so where's the white rabbit?', it didn't seem worth replying, 'I'm named after Alice Springs.'

Australia didn't entice me. Although my grandfather had only swapped Adelaide for Cambridge at sixteen, and a great-uncle had been killed fighting at Gallipoli, my family didn't see themselves as fifth generation Australians. In fact, the Antipodean interlude was regarded as a blip. The southern sun hadn't changed the gene pool. We didn't return with bulging muscles and tawny skin. The women in our family still tended to marry absent-minded scientists, and Cambridge was once again our home town. Only once did my grandfather, by then a Nobel Prize winner in physics, reveal his roots. When a rural dean asked him if he was descended from convicts, he hit him.

A letter once arrived from Alice Springs. Their tourist board had tracked me down and wanted me to attend some celebrations and appear on a float dressed as the young Mrs Todd. At fifteen, I was prepared to wear anything outrageous, but after reviewing the photograph of the austere lady, I opted to go on a French exchange instead.

During university summers, I travelled to India and South America and made my way through the Far East from Tibet to Hong Kong. As a journalist I was sent to Eastern Europe, America and South Africa. Only once did I touch down in Australia and that was for less than twenty-four hours. I was with a group of travel writers on a ludicrous quest to circle the globe in fourteen days, visiting as many countries as we could, to review a new round-the-world ticket. In each place we had to do one activity. Australia was scuba diving. We queued up with hundreds of other penguin-suited tourists to jump off a ship over the Great Barrier Reef.

As a wedding present, a relative sent me two more pictures of Alice. One was of her as a child, in black silhouette, as she played alongside her ten brothers and sisters. The other was a photograph of a sixty-year-old white-haired woman looking wistful, with watery eyes and a kindly, haunting stare. Neither bore any relation to the straitlaced matron I had seen before. I was also given a small bunch of barely decipherable letters from Charles to his wife, and my mother gave me Alice's silver pickle jars.

Then a distant Australian relative sent an old scrapbook with pasted newspaper cuttings and a note saying, 'found these stories in the attic, thought a surviving Alice might like them'. Entitled 'Telegraph Todd', the series of articles had been written in the 1950s for Adelaide's *Chronicle* by Lorna Todd, Alice and Charles's youngest child.

I began to wonder why this apparently treasured Victorian child was prepared to put half the world between herself and her comfortable family life, especially as she must have been warned by chattering Cambridge that Australia was not a place for young ladies. The only clue I had was her proposal, which implied that she was either wildly spontaneous or that there was some compelling need for her to escape. She didn't seem in love with this strange, intense man when she rescued him from his life as a bachelor. Did she ever fall in love with him?

Charles's motives were easier to fathom. He was absorbed in trying to criss-cross the new continent with telegraph poles. His life, I was told, was etched out in yellowing documents held in the state library and by Australia's telephone company, Telstra. There was a 1955 Boy's Own adventure story called *Overland Telegraph* by Frank Clune about Todd's escapades and a half finished biography from the 1970s by a Major-General G. W. Symes. But the man himself remained remote.

I started cornering my mother's older relatives at weddings, christenings and funerals to try to decipher some of the myths. One branch thought that Charles gambled everything in a gold boom and lost the family money. Another cousin was convinced that Alice's sister, Sarah, had become pregnant out of wedlock, and the family scandal ensured that Alice was packed off as quickly as possible. The most obscure was the story I heard about Alice's great-grandmother, who at twenty went to do the season in London, came under the influence of a famous Congregational preacher, and eloped with him one night after clambering out of her window.

Fact and fiction needed untangling. As Charles would have pointed out, a message can only be transmitted along a straight wire. The slightest distraction, a mere cobweb, risks diverting the dots and dashes down the poles and into the earth. The stories I was receiving were frustratingly garbled. Occasionally, I would piece together small chunks of my great-great-grandparents' lives. Then the line would go dead.

I decided to go to Greenwich to trace Charles's family. Charles Heavitree Todd was born on 7 July 1826 in Islington, the second son of a Welshman named Griffith Todd and his wife, Mary Parker. Griffith was a melancholy character who, despite being a Dissenter, drank to forget his problems. He was variously described as a grocer, dealer in wines, and tea merchant. By the time Charles was seven, his father was floundering. Griffith, unlike his son, had no head for figures and was too whimsical for trade. The Todds moved to Greenwich, where the family scratched a living selling 'English

Wines' from a shop in Church Street, which just paid the rent on a small house, No. 78 Ashburnham Grove.

Charles's mother was referred to as 'the invalid' by her son. She would spend her time in bed or at the front window of the parlour reading religious tracts. She came from a strong Dissenting family and prayers before breakfast would drag on for nearly an hour. The three sons went to the local Roan school. Griffith George, the eldest and most gregarious, left at fifteen to become a pilot on the Bengal coast. Charles's sisters, Mary and Elizabeth, became strict Nonconformists like their mother. Mary was an exceptional pianist, who once played before Queen Victoria. The youngest brother, Henry, later married one of Alice's cousins.

In letters and diaries, Charles barely mentions his childhood. The only anecdote he passed on to his children was that occasionally he would play at being parson, wrap up in one of his father's nightshirts, put an upturned washstand over his head and preach until even his mother was helpless with laughter. As he got older, he didn't disown his roots. He was protective of his younger siblings, and secured Henry a job at the Observatory. Perhaps feeling guilty that he had escaped to the other side of the world, he wrote them long letters about his new life.

When Charles first went to the Observatory, built by Sir Christopher Wren in 1675, it had become so dilapidated that the First Lord of the Admiralty suggested razing it to the ground. No one had bothered to analyse the mass of observations of the moon and the planets accumulated over the past ninety years. Sir George Airy managed to persuade parliament to give him a grant of £190 to reorganise the department and employ more human calculators. Charles was the last of the six to be chosen, on 16 December 1841.

As a supernumerary computer, he had to work fifteen hours a day in ill-lit rooms, with only one break for dinner. There was a minor gap in his service, presumably through illness, but this didn't prevent him becoming an assistant at the Cambridge University Observatory, the highest position any public servant could expect to attain without a formal education. His letter of commendation calls him 'a faithful and useful assistant, a good observer as well as a good and quick computer'. The annual reports make it clear that Charles spent hours following the movements of the stars and comets through Cambridge's new Northumberland telescope. His little spare time seems to have been spent at the Bells'. He was listed as being at their house during the count for the 1851 census, although Alice was elsewhere. In May 1854, he returned to Greenwich as an assistant in charge of the time-balls.

Charles almost didn't go to Australia. Sir George Airy had first recommended a better connected young man who declined, going instead to a more prestigious posting in Canada. In 1855, South Australia was only a fledgling colony, almost twenty years old, but it desperately needed more public servants. The Crimean War, in 1854, which pitted Britain and her far-flung empire against the might of Russia, had made the outpost feel vulnerable. The colony wanted a telegraph link built from Adelaide to its port so that some warning could be given if the Tsar's fleet sailed up the coast. This was hardly likely, but their isolation bred exaggerated fears.

The South Australians asked the Colonial Office in London to find a qualified man to take charge of astronomical observations and to set up a telegraph line. Sir George was concerned that his reticent pupil would wilt in the Australian sun. He wrote to the Colonial Office, 'I doubt whether Todd has the boldness and independence of character which may be required in an Australian establishment.' But he couldn't think of anyone except Todd who knew about astronomy and telegraph systems. So he set the twenty-eight-year-old a test. The time-ball at Deal had developed a fault. It needed to be repaired urgently. Could Todd do it alone? He did so within a week and was offered the job on 10 February 1855. Charles didn't hesitate before accepting it, although according to family stories he did mumble something about having to visit a young lady first.

My next trip was to Cambridge. I dug through the county archives, which confirmed that Alice was born on 7 August 1836, and had ten brothers and sisters. Her father was a cereal and seed merchant, who married a vivacious bonnet-maker six years his junior after meeting her at a horse fair. By the time Alice was born, they were considered prosperous. Alice was taught the piano by a series of governesses and music teachers, and also learnt to dance and write a good hand. Simple arithmetic was the only exercise she loathed. A bright, engaging child, she was indulged by her elder brothers and sisters. She had a soft Cambridgeshire accent, having never travelled outside the county, but she could also speak French. The family were active in the Free Church, and civic duties. After she left, Alice's brother Edward sold the house in Free School Lane and moved to a manor house in the nearby village of Chesterton. He eventually became mayor of Cambridge, as a reward for which he merited a long obituary in the *Cambridgeshire Times*.

I could find little else on Alice. Then a Todd descendant sent me a family tree. Under Charlotte and Edward Bell were the names of their children – William, Sarah, Edward, James, John Gillam, Charlotte, Alfred, Alice, Charles, Elizabeth and Henry – and under that their dates. Seven of Alice's brothers and sisters had died before they reached twenty-one. The silhouette must have been commissioned the only year that all eleven children were alive. William, the eldest, and Henry, the youngest, died when Alice was only seven. A year later, John Gillam died. Eliza died when Alice was nine, James when she was eleven. So Alice lost five siblings in the five years before she first made her rash proposal to Charles Todd. Her closest sister, Charlotte, with whom she had always shared a bed, died when Alice was fifteen. By the time Charles came to tea when Alice was eighteen, there were only four Bell children left: Alice, Sarah, Edward and Alfred.

How did they all die? It seems that some of them developed a form of tuberculosis. Alice's childhood may have sounded a happy one, but the family home must have become increasingly oppressive. No wonder the twelve-year-old Alice thought of escaping if five of her family had just died. Henry had been only one, Eliza and Charles five, old enough to have become playmates for Alice. William and James both died at nineteen, just as their father hoped they would begin to make a mark on the world. Infant mortality was high in the mid-nineteenth century. But Alice's aunt Sarah had nine children, all of whom survived their teens. For Mrs Bell, the loss of seven children must have been an extraordinary burden to bear. Alice's sister Sarah remained a spinster at home to help nurse her family, so great hopes were pinned on Alice, the youngest surviving child, and clearly her mother's favourite. Losing her to Australia must have been a harsh blow for Mrs Bell, but she seems to have accepted her daughter's decision with remarkable sanguinity. She might even have felt that the warmer climate would be healthier for her daughter. Although Alice was close enough to Sarah to give her the inscribed *Pilgrim's Progress* before she left, and donate to Alfred her butterfly collection, she was clearly determined to put some distance between herself and her early, troubled life, marked by black mourning dresses.

On the day I visited, the church where Alice and Charles were married was holding a Korean service, and an electric piano was belting out unfamiliar tunes. Alice's old home in Free School Lane was still standing, opposite a lingerie shop and next to a Greek restaurant. The seventeenth-century house looked poky in comparison with the gleaming zoological buildings now standing on its other side. The house where Charles acquired a wife had become a transit lounge for visiting academics.

Despite finding a few more letters, the trail had gone dead, but I was now obsessed by this fragment of family history. So I decided to follow Charles and Alice out to Australia. It remained only to convince my husband, Ed, that he needed a holiday in the southern sun.

3

... — —

Crossing the Equator

Charles, in 1855, had four months to prepare for the new continent, and he used the time to talk to every telegraph company in the country. He quizzed returning colonists about the availability of mechanics and labourers, whether there was timber for telegraph poles and what climatic conditions he could expect. Sir George Airy personally helped him to choose the telescopes and chronometers that would be necessary to set up a self-sufficient observatory. Although the colony had asked for four men, Charles decided to take only one assistant with him, Edward Cracknell, five years younger than himself, who was employed as an instrument maker by a London firm.

Almost a hundred and fifty years later, my preparation amounted to booking two return tickets to Adelaide from the back of a Sunday newspaper supplement a week before we left and rooting out the addresses of my remaining antipodean cousins. As we scrunched up in our cramped charter aircraft, we started deciphering Todd's sixteen-page letter to his family about his boat trip across the world. Stranded in the chilly air-conditioned terminal at Kuala Lumpur, waiting nine hours for a late connection, we ate Dunkin' Donuts and dried lychees and read out to each other the descriptions of the feasts which the Todds consumed on their voyage. We fantasised about the five-course meals they dined on at the captain's table. Breakfast alone consisted of 'splendid ham, calf's head, fresh herrings, salmon, cod, beef, tongue, hot rolls and honey'.

There were five pigs, nine sheep and several chickens living in the ship's larder when the *Irene* set sail. The cabbages lasted until the middle of September, but the carrots ran out after the Bay of Biscay. They dined at three, and at six had a tea of bread and preserved gingers. Before bed, they wore all served a measure of grog. Todd mentions being becalmed in the doldrums and nearly killed in the Roaring Forties, but he complained more about his indigestion and the fact that the cook's fingernails were always grubby. With their jovial skipper Captain Bruce for company instead of our surly flight attendants, we were unsympathetic.

Charles probably realised how lucky he was to be sailing on a ship of willing recruits for Australia. As a child, he might have seen the packed convict transports waiting on the Thames to set sail for Sydney. The prostitutes who hung around on the street corners would have known they risked being rounded up and transported in cramped, stinking holds to the other side of the world. Until the 1840s, Australia had been the great dumping ground for 160,000 'criminals'. By contrast, the Todds' trip was so sociable – playing games like shuffle-board, a mature form of hopscotch, with the other passengers – that Charles admitted to having little time to study. 'In the stern cabins there is almost always too much motion, and in the cuddy too much talk,' he explained in a letter which he dated 'Latitude N2 53S, Longitude 115NNE'. Charles spent two hours a day practising with his instruments, but whenever he tried to plot the ship's course, the captain always beat him to it.

The sixteen other passengers included an alchemist, a baker, a cleric and a gold prospector, who was convinced he would find a nugget in the outback big enough to pay his debts. Six returning colonists regaled them with tales of bush heat, fleas and hot winds that provoked the strangest passions. They warned the Todds of fires that burnt a path fifty miles long and twenty miles wide, one of which had destroyed 1,400 sheep, a shepherd and a dog. Charles was undeterred, writing that these fellow passengers 'were now returning to Adelaide ... being unable to endure the changeable climate of England after that of South Australia'. Increasingly optimistic about his new posting, he added 'the climate by all accounts is very fine and fruitful'.

They whiled away the hours singing ballads on the deck and counting flying fish, albatross and mollyhawks. The men caught jellyfish to see if they glowed in the dark and as they got closer to Adelaide they spotted Trinidad pheasants, Cape hens and whale birds. They would try to catch birds by putting a piece of fat pork on a hook attached to a line and floating it astern, or by shooting them. The albatross was their prize target, 'the king of seabirds larger than a swan', according to Charles, who never managed to shoot one. The birds would circle the boat for hours or dive-bomb the passengers. 'It is the pace with which the albatross fly,' Charles remarked, in a rare poetic moment. 'They scarcely move their wings, yet they soar and sweep gracefully airwards almost majestically.'

The new government astronomer gave impromptu astronomical discourses and officiated as chaplain at the daily prayers. Happy to find himself looked up to for the first time, Charles couldn't help boasting, 'I have thrashed all at draughts till no one will play,' and adding, 'there are several chess players on board, but only one with whom there is much pleasure in playing, the others are novices.'

The Todds must have made an odd couple. Not only was Charles a good three inches shorter than Alice, but they also seemed distinctly nervous in each other's company. Charles was alarmed to discover that while Alice may have been equipped with dozens of newly monogrammed sheets, she was, in other ways, totally unprepared for marriage. 'My mother was the youngest of a large family,' her daughter Lorna wrote. 'She was brought up very strictly, seldom out of the nursery and schoolroom. She married from a household well staffed with servants, knowing nothing of housekeeping or indeed of life.'

The old maids of Cambridge had explained to Alice that her

complexion would wither in the sun, and her brothers had told her that she would have to live off whale meat, but no one had given her any practical advice. Lorna writes that her mother used to entertain her teenage daughter with stories of that first trip, hinting at how ignorant she was of the role that Charles might expect her to play in the bedroom.

The new bride and bridegroom were never alone until they set sail. Even then, Alice had her own cabin, which she shared with Eliza. For the first few days on the boat, Eliza felt so queasy that she had to sleep out on the deck. Alice was scarcely more comfortable in her bunk. She was unused to sleeping alone, and she locked the door in case the deckhands got too rowdy over their cards. In the next-door cabin, Charles spent several nights debating what to do. It seemed easy to link hands as they went in to dinner with the captain, but he didn't know how to progress any further.

The problem was solved by the first storm, which pitched his wife out of her bunk. She landed in a pool of water and injured her shoulder. Hearing his wife sobbing, Charles lost all inhibitions. He summoned the steward and demanded that he open the door. Alice must stay in his bunk so he could anchor her in, he said firmly, and lifting up his sodden wife, he deposited her on his armchair.

'Are you homesick?' Charles asked.

'No,' Alice replied. 'You see I was frightened of Mama.'

'And are you frightened of me?' Charles queried.

'A little, you see I don't know you very well, Mr Todd.'

Charles found Eliza and told her to bring the copper warming pan, adding firmly, 'I will keep her here with me for the night.'

According to Lorna, who knew the old maid well, Eliza retorted, 'If I hadn't felt so bad, I'd have suggested it before.' She went off to find a hot posset. As Alice thanked Mr Todd, he told her firmly that, from now on, she was to call him Charles.

From that night, the trip passed more smoothly. The couple were obviously soon better acquainted, because Charles writes home happily about their sleeping arrangements. 'It blew hard on Monday and Tuesday so we were obliged to have our mattress on the floor.' He had also nicknamed his wife 'Lal.'

They soon hit more storms, with waves reaching thirty-six feet. 'Sometimes one gets a cup of scalding tea on the head as I did,' Charles wrote. 'Others have ended up with gruel streaming down their faces.' The passengers would be sitting at dinner with both cuddy doors shut, the meat carved and the potatoes served, when a wave would hit the ship, dashing open both doors and leaving them drenched in water. Or they would be sipping their turtle soup and the tureen would be knocked sideways, scalding their legs while their feet would be immersed in a freezing puddle of water.

Early one morning, Alice heard a crash followed by a scream from the man at the wheel. The second mate skidded past and started yelling for help. The captain, who had a bad limp, rushed on deck. The chain that linked the wheel to the rudder had snapped. The vessel, with all sails up, was running out of control at thirteen knots. Everyone waited for the ship to flip round and the masts to crack. Charles inched his way forward to see the captain 'as pale as death' giving orders and men clambering over the sides. The rudder was finally connected by threading a rope along a pulley and sliding it down the bulwarks of each side.

The newlyweds were more disturbed by the doldrums than the storms. These, Charles explained, were a belt of windless seas lying near the equator. After three weeks becalmed, Alice panicked, fearing that they would be stranded for ever between Cambridge and Adelaide. Her husband, beginning to realise just how young his wife was, explained that she too was in the doldrums, a psychological as well as geographical and meteorological state. To keep her entertained he spent the evenings on deck identifying the stars. Together they would wait for the first sighting of the Southern Cross and Magellan's Clouds. He showed her Venus before sunset and Jupiter at sunset, and they could see Mercury without the aid of a telescope.

At the end of September there was a shout of 'land ho'. They had sighted Gough Island in the South Atlantic. Charles, after consulting his books, was triumphant. By working out their longitude and latitude, he had discovered that the captain's chronometer was thirty seconds slow. The island was used by whalers for boiling blubber and they could just make out the fires. For the restless Todds, 'even this distant view of land did us landsmen some good'.

After two months, Charles finally plucked up the courage to tell his wife more about his grand plan, only to discover that Alice had no time for the mechanics of the telegraph. When he explained that he wanted to form a metal girdle, like her crinolines, round the world, which would enable everyone to communicate with the motherland, she became more interested. Charles explained that he might have to spend long periods away from home plotting the routes for his telegraph poles, but she would soon have children to keep her company. Alice insisted she was happy with her cat, which she had brought from Cambridge.

As they rounded the Cape of Good Hope, Alice felt sick every morning. The boat records don't detail a stopover in Cape Town, but Lorna insists that her mother talked about disembarking there, and Charles, worried about her health, found her fresh milk. In the next few weeks, as Alice rallied, Charles was convinced he had married the right woman. His new wife was beginning to show pioneering mettle. She talked less of her family and Cambridge gossip, and more about the animals they would find in the new country. Todd's assistant, Cracknell, had brought his wife, who became firm friends with Alice. When Mrs Cracknell's baby died 'of starvation' Alice lined a box with one of her capes as a coffin and organised a funeral. Everyone assembled as the box was laid on a board and covered with a Union Jack. Alice played her piano and the crew looked out for sharks as the coffin slid into the sea.

Alice was turning into a spirited young wife and the centre of the ship's social life. When the temperatures plunged, she didn't mind wrapping up in her furs, although Todd kept grumbling about his chilblains. 'The voyage has done us a great deal of good, especially Alice,' Charles wrote home. 'She suffered from sickness in the first months but since then she has got quite stout and hearty. I was only sick twice. Alice's spirits are very good, and by general kindness she is favourite with all, and a great comfort to me.' Eliza became the butt of their jokes, always worrying that the boat was about to tip over. When they reached the equator she was told to listen carefully as the vessel scraped against the line. 'The men forward had a little spree, some passengers sat up and a few headaches resulted,' Charles reported.

The captain promised the last of the mutton to the first person who sighted Kangaroo Island, ten miles off the coast of South Australia. Cracknell claimed the prize four-and-a-half months after they set sail. The island reminded Todd of the headland off Deal. He thought it 'very picturesque', and laughed when Alice said she could smell the seaweed on the coast. Indeed, she insisted that it was just like going on a day-trip to Felixstowe.

Eliza cleaned the cuddy, the male passengers helped to paint the poop, and Todd worked out that Adelaide was 9 hours, 14 minutes and 20.3 seconds ahead of Greenwich Mean Time. They were becalmed for two days off the coast. As they began to move slowly from Cape Jervis, the gum trees rose up along the Australian shoreline. But there wasn't enough wind to make it over the seventeen foot sandbars at the entrance to Glenelg for nearly a week. An outgoing ship gave them a newspaper which Charles picked up with relish, only to discover that the news from Britain was dated two days after they had left Dover. 'A long voyage seems like a blank in one's life,' wrote Todd, 'so in a way it appeared quite natural to have news from England but a few days after our departure.'

A few of the men, including Cracknell, took a small boat ashore and rode into town. Todd decided the journey was too expensive. He spent the time teaching Alice to fish. By the next day she had got the knack and was pulling up dozens of garfish, the piano was tuned and everyone 'felt jolly'. In the evening there was a tremendous thunderstorm. 'Most sublimely grand', according to Todd. Ten days later, Charles was greatly excited when he met Dr Rankine, a member of the Legislative Council, who had lost ninety-four sheep in the storm. For an amateur meteorologist like Charles, this land had potential, and he was soon addicted to the extremes of Australian weather.

As they came up the coast, Alice could see nothing but flat plains through her husband's binoculars, but she was used to that from the Fens. She was disappointed that the landscape seemed so similar to Cambridge-shire. The boat was finally tugged into the port at Glenelg on 4 November 1855. They were greeted by the sight of a large grand piano, a four-poster bed and a wardrobe stranded on the sand. 'Glad now you only had a cottage piano', Charles laughed. 'That furniture is too big for any bullock wagon.'

It was late spring when they landed, and Alice could find neither the jungle nor the desert she had been warned about, just grass. She swallowed her first antipodean insect while struggling into her red flannel petticoat, her two white petticoats, tight stays and new crinoline dress. Now that she was married, she had to wear a matron's bonnet on her head with wide satin ribbons. Eliza was determined that her mistress should look the part. Alice, already sweltering, refused to wear the elastic-sided boots and chose instead her impractical black satin sandals.

Only on arriving in Glenelg did Alice appear as flustered as I was when approaching this new continent. Her sideboard got stuck in the gangplank in much the same way that my tennis racket managed to snarl up the conveyor belt at the airport. But unlike us, she wasn't pulled aside by customs men and searched in case she was trying to smuggle orange peel, seeds or any other foreign substance into the country. On her ship, Captain Bruce had encouraged them to bring crocus bulbs, foxes for hunting and pet rabbits, all the ingredients for a new England. Alice's cat may have had the invidious effect of helping to swell present-day Australia's detested feral-cat popula-tion. No one on her boat dreamt that the animals they were transporting might cause havoc with the continent's ecosystem. Yet less than a hundred years later those pets had forced modern Australians into biological warfare against the immigrant rabbit population, using the myxomatosis virus.

The next day, there was a grand regatta at Glenelg and a holiday because it was 5 November, Guy Fawkes Day. The passengers watched a steamer taking gaily dressed holidaymakers out on the river, but by then they themselves had had enough of the water. Charles presented Captain Bruce with a silver cup on behalf of the passengers, and the Todds set out by bullock wagon for Woodville, three miles outside Adelaide. They wanted to visit some family friends, the Halls, who had recently emigrated from Cambridge. Once again, Eliza was ill, this time from the dust and the flies, and Charles began to worry whether she would be an appropriate companion for Alice while he was away.

Mrs Hall was recuperating from a chest infection and was in bed when they arrived. The embarrassed Todds explained that they had only dropped in to pass on home news. But as they drank tea, a dust storm began to blow. 'This could last for hours,' said Mr Hall blithely. The doors were closed, but the rooms still filled with fine dust. Alice thought she was suffocating. When she went to the window she couldn't see more than a foot outside. The Todds had no choice but to stay the night.

The house had four rooms. Alice was appalled when the old couple insisted on giving up their bed and sleeping on the parlour floor so the Todds could enjoy a comfortable first night. It seemed rude to leave too abruptly the next morning and soon the visit had stretched to a week. The house was poky and noisy, and Alice struggled to remember her pioneer spirit. Eliza was more unhappy. She had to sleep in a wattle-and-daub cottage next door and grumbled that she had been 'half et up with mosquitoes'. She never got used to the continent and was called Poor Eliza for the rest of her life.

Mrs Hall was an acknowledged saint, which soon grated on Alice's nerves. Her neighbours were constantly telling Alice about her little acts of kindness. Everyone knew she was sleeping on the floor, even though it made her rheumatism worse. Her husband had an unnerving habit of coming up behind Alice as she rested on the veranda, clutching her shoulders, and whispering, 'You'll make a wife yet.'

Alice pleaded with Charles to let them stay in a guesthouse to relieve

Mrs Hall, so they moved to Miss Faulkner's boarding establishment, although Charles fretted about the weekly rent. The Todds were never allowed to forget the Halls' charity: 'As a family we suffered unduly for that act of kindness,' Lorna wrote to a cousin fifty years later.

After their first fortnight, when they were well infested with fleas, a letter came saying that the government had managed to secure a small two-storey house for the new public servant in Sturt Street, such a socially dubious location that Lorna later claimed that her parents had lived in Angas Street. Alice didn't care that she would be living in a wooden cottage on the wrong side of town. She wanted to play house with her new silver. Nor did she mind when Mrs Hall warned her that there would be no running water and she would have to buy supplies from a cart that passed every day with barrels from the Torrens River. All she wanted to do was to take off her new stays, and be able to strip to her undergarments and bare feet without any prying eyes.

Todd later wrote to his parents that they would soon move into 'a six-roomed house of two storeys on Brougham Lane, a wonder here where most are all on the ground floor'. He added proudly, 'The bluestone Observatory is to be commenced immediately. I shall then have a beautiful house, six rooms on the ground floor with calculating rooms, a nice veranda front and an acre of ground.' Mrs Hall fussed around giving Alice advice. She insisted on donating her canvas water bag to hang in the shade and a calico bag for meat and butter. She told Alice to drink only boiled tea. Charles, the tea merchant's son, had bought a large supply of his Orange Pekoe brew. Finally they were off to the city.

Their first view of Adelaide was disappointing. Built on a rich alluvial and limestone plain, and named after the wife of King William IV, the city had been designed on a grand scale. But when Alice arrived there was just a straggle of shops with wide verandas and no glass in the windows. Only two or three buildings reached three storeys, the roads were unmade and there were no footpaths. Horses were tethered anywhere, bullock wagons outnumbered traps and the rubbish swirled at the street corners.

Alice felt stifled by the scorching winds, but Charles found Adelaide's bizarre weather patterns intoxicating. The hot north winds were the most extraordinary potent blasts which would hit the cool sea breezes, causing the clouds to pile overhead and sit on the city for hours. The new house contrived to be damp even in the heat. After a week, Alice received her first calling cards. They were home-made affairs, but signs at least of recognisable Victorian society, and marked the real end of Alice's childhood as she struggled to establish her social position in the new colony. She was reassured by her neighbours that Adelaide was different from the rest of Australia. Although convict boats had supplied the other colonies, no immigrants to Adelaide had secured their passage by stealing a pheasant. South Australia, founded in 1836, was known as the paradise of dissent and had been conceived by philanthropists as a mini-Utopia for religious groups seeking tolerance. The philanthropists hoped that the colony would become 'an extension of the old country with all the good but without the evils of the old society', according to evidence they gave to parliament, and they aspired to create 'an earthly paradise of perfected human nature'.

The colony was soon home to various religious sects. The German Lutherans, unwanted in Frankfurt, had settled in the Barossa Valley, unrolling their beach towels on today's prime wine-making country. The Quakers were helping to build a library. The Protestant Italians were making dried kangaroo taste like prosciutto. This was the only town in Australia that could boast more churches than pubs.

Unlike America, the Lucky Country had no bewigged founding fathers with their tablets of good intentions. Mark Twain, visiting from the older colony, dismissed Australia as an 'entire continent peopled by the lower orders'. But the South prided itself on being a cut above the rest, even though it could never compete with the commercial acumen of Melbourne or the new money of Sydney.

Anthony Trollope immediately saw Adelaide's potential when he visited in the 1870s. A postmaster himself, he is likely to have met Charles while touring the imposing new postal headquarters, but his journal of the trip makes no mention of a Mr Todd. 'No city in Australia gives one more fixedly the idea that Australian colonisation has been a success,' he wrote. Even so, South Australia was still volatile enough to have forty-one governments in its first thirty-six years of parliamentary rule.

The new country suited Todd. In England, he had felt awkward with his thin frame. His family teased him for being a child prodigy and his neighbours tried to get him to use his mathematical talents to pick odds on the horses. In Cambridge, he was painfully aware that his colleagues were university educated. But the moment Charles met the Governor in South Australia he felt at home. 'I am, I can assure you, quite an aristocrat here,' he half joked. He had standing and, equally important, space for his experiments.

Charles was undeterred by the fact that a private telegraph system had started the day he arrived in Adelaide, linking the city with the port. Within a month, he had set about building his own government line and put the cowboy operators out of business. Within a year, his new enterprise had transmitted 14,736 messages yielding a revenue of £366. He was beginning to pay his way.

The newly married couple were on the rota for flowers in the church, the focal point of all social gatherings. Life was basic, but the frugal former bachelor didn't mind. They bought chickens and a cow and their evening entertainment was Alice playing her piano. When Alice was heavily pregnant, they moved to their grander premises in Brougham Lane, with a garden and fruit trees. Four months after their arrival, Charlotte Elizabeth was born, on 12 March 1856. The baby, whom Alice named after her elder sister, became known as Lizzie. She must have been conceived shortly after Alice fell out of her bunk.

The teenage mother amused herself learning to be a wife. She read out snippets from the newspapers over breakfast and found novel uses for her wedding presents, such as the silver asparagus holders which she used to clip back the fly nets. At first, Charles insisted on accompanying Alice and Eliza shopping, worrying that 'most things are excessively dear' and he must teach them to be prudent. Bread was 1s. for a two-pound loaf. But lamb was cheap and the vegetables were good.

Alice's only complaint was that her husband would often become so engrossed in his scientific ramblings that he would forget that she was there and let his fried eggs congeal on the plate. But life in the new continent never seemed dull. After six months, the Todds managed to hold their first dinner party. The new cook was so clumsy that she cut her hand while making the creamed potatoes, turned them pink, and tried to pass them off as flavoured with beetroot. Staff in Adelaide were a problem, as the editor of the local newspaper, the *Pastoral Review*, pointed out: 'Unfortunately four fifths of our servants are liars and dirty.' Any decent woman was soon snapped up as a wife.

Todd was envied for having married such an enterprising woman. She may not have been a great beauty – her nose was slightly too wide, her mouth too slight – but she was admired for her elegant style, she had genuine charm and was enthusiastic, energetic and healthy. As a couple they received more invitations than the bachelor Todd could have hoped for, and were soon taken under the wing of the Governor and his wife. Their first year was a happy one. Alice's gamble seemed to have paid off.

4

·····—

An Upright Metropolis

One hundred and forty-two years later, our first night in Adelaide was spent in the Hyatt hotel trying to turn off the air conditioning and deciding whether to sample a charming local delicacy, a pie floater, from a van outside. The van attendant explained that the large meat pie he was showing us would be floated on top of a warm mushy pea soup and drenched in tomato sauce. The art was in getting the pastry light enough so the pie was suspended in a sea of grey-green pea. We decided to see what was in the mini-bar instead.

It was Sunday night, and Australia hadn't yet grabbed us. Then the phone rang and our first contact said she was downstairs ready to drink daiquiris. Jane was from the South Australian tourist board and had brought along her partner, Robin. Both women had matching turquoise-feathered earrings, short cochineal hair, pixy faces and expansive laughs.

Australia's deep south is still nothing like the rest of the continent, they said. Adelaide retains a haughty disdain for the brashness of Sydney and Melbourne and the relaxed banter of Darwin and Perth. They return the compliment by insisting that Adelaide is atypical of the great Aussie experience, as snobby and stuck up as the old country. Although half of

Australia's inhabitants have at least one parent born outside the continent, old Adelaidian families pride themselves on being seventh generation antipodeans. Ostensibly, the city is a stultifying morass of etiquette. People sit on their wooden verandas talking about the English county cricket scores. There are knitting circles, children wear straw boaters, and old ladies with pastel gloves and matching handbags hold fund-raising tea parties in the botanic gardens.

But within days, this upright metropolis seduced me. Planned in the nineteenth century by Colonel Light, the Far East's equivalent of Baron Haussmann who remodelled Paris for Napoleon III, it has kept its fine bone structure, despite wrinkles appearing in places. Framed by the Mount Lofty Range and bisected by the Torrens River, a series of gracious squares and boulevards are set in a figure of eight, with the cathedral in north Adelaide, and the commercial district at the bottom of the hill in the southern loop. The city is surrounded by parkland where daughters ride ponies and grandfathers play golf. Black swans swim up the river past the rowing clubs. Genteel Adelaidians send their sons to 'public schools' with cricket pavilions. Much of middle-class Adelaide came over from Britain to escape the depressing 1950s. They were enticed from their terraced houses by £10 one-way tickets and were known as the '£10 Poms'. After thirty years, many have earned their swimming pools, tennis courts and bridge afternoons.

Yet Adelaide has also acquired a reputation as the most liberal place in Australia, stemming possibly from its tolerance of any religion. In the nineteenth century there was a strong temperance league, but also plenty of drinking. A hundred years later, the state had Australia's most flamboyant Premier, Don Dunstan, who wore pink shorts to the office. South Australia produced the first openly gay chief justice and the first nudist beach. Most of our friends seemed to be gay. I couldn't see how the city managed to repopulate itself.

South Australia was also the second place in the world to allow women the vote, and remains zealously politically correct. Chocolate bunnies are off the Easter menu because they are not an indigenous species. Instead, we celebrated the festival by eating a long-eared chocolate rat-like animal called a bilby, a native of the hills outside Adelaide. Students in tie-dyed baggy trousers work in feminist bookstores or train to be chefs, and the city is at the forefront of modern Pacific-Rim cooking. It also has a seamier side; other Australians whisper tales of murky paedophile rings, which they claim have been covered up by the old Adelaide establishment. The city appears to be suffering from advanced schizophrenia, and even hired a town psychiatrist to diagnose what was wrong. After spending £1.4 million, he reported that Adelaide was the most perfectly designed city in the world; more eccentric old lady than dangerous psychopath.

Presiding regally over the alternative establishment was Don Dunstan, who had retired as Premier and now owned a local restaurant called Don's Table. We went to see if he could explain this unusual paradise. Don, in his seventies, was wearing cut-off shorts. His Chinese friend made us tea in their wooden chalet, festooned with dusty tropical plants. Adelaide in the sixties became Don's playground. Until he died, in 1999, the former Labor Premier was still fighting old political battles between stir-fries. Dressed in his safari suits, with a glass of wine in his hand, he dragged Adelaide into the twenty-first century, and started the crusade to procure land rights for the Aboriginal people. He both shocked and thrilled South Australia. His enemies were the white pastoralists beyond the city limits, living the traditional life in the saddle, and what he called the OAFs, the old Australian families.

On our second day, we set out to find Charles and Alice's first home, but instead we stumbled across their church near the post office with a brass plaque on the wall to the two Todds. A man approached us as I copied down the inscription on their memorial. He had been a surveyor for the new railway line to Darwin, and when I told him my connection he said that the company he worked for had used Todd's maps from the last century and offered to show them to us.

Embarrassed, I explained that I didn't know much about Todd's life. So we sat on a pew and he explained that my great-great-grandfather had been obsessed by opening up the red continent. He took us to his office to see photocopies of Todd's sketches. The surveyor was a large man, but extraordinarily delicate with the maps. 'We used to spend hours camped out in the bush at night wondering how the old buggers had managed to be so accurate with their measurements,' he said. 'They could calculate everything by the stars, down to the last foot.' The maps had Todd's delicate writing scrawled all over them.

After the first year, Charles was frequently gone for months on end. Only days after Lizzie was born in 1856, he was drawing up travel plans. Encouraged by the success of his first telegraph line, he agreed with the new Governor, Sir Richard MacDonnell, that they should build a connection to Melbourne, linking the two colonial capitals. This, he accepted, was far more pressing than a detailed study of the stars.

The Governor dispatched the young man by ship to haggle a price. There Charles met Samuel McGowan, Victoria's Superintendent of Telegraphs, who became his lifelong friend. McGowan was the first man to build a telegraph line in Australia. A Canadian, he had studied under Samuel Morse before coming south and introducing his professor's system to the new continent. The two men agreed that the Victorian line should be extended to the border and Charles should build a three-hundred-mile stretch to meet it. Charles bought a horse from McGowan and made his first outback trip back to Adelaide, over six hundred miles, learning to ride along the way.

As the line progressed, Charles would ride along the isolated route encouraging his teams. One evening, he took a detour and came across a man in a limestone cottage boiling his billy can. Charles asked to share his brew. On discovering that the man was an Oxford MA, he challenged him to a game of chess, and greatly enjoyed beating him. All his life, the Greenwich boy took pleasure in demonstrating that his lack of a university education was no reflection on his intelligence. When Charles tried to look him up on his next excursion, the man had fled, leaving a note saying that this stretch of country was getting too busy for his liking.

The line was finished in July 1858, and three months later Melbourne and Sydney were also joined. So now the three capitals could communicate within minutes and messages of 8,000 words soon became common. A flushed Charles was awarded £1,820 good-service pay. He was also praised for having discovered that the position of the 141st meridian, the boundary between South Australia and Victoria, was two and a quarter miles out, and by moving the boundary in the South's favour he acquired much needed new arable land for the poorer colony.

Charles adored the open spaces of the bush from the moment he learnt to ride, and after that first year he found every excuse to return to the scrub. Lorna said he never let the children read a book in a railway carriage. 'The scenery may not be beautiful,' he would say, 'but it's strange and it's new.'

There is little evidence of how Alice fared when her husband was away. With none of her family to keep her company, and few women of her own age for friends, she must have felt isolated as she struggled to nurse her first child. Mrs Hall and the other matrons fussed around, insisting she take her responsibilities seriously. But she had to make the transition from childhood to motherhood in only a few brief months. What was once seen as charming spontaneity was now dismissed as recklessness. Pioneer life, with its endless round of cooking, cleaning and making do with coarse flour and rough-hewn furniture, must have begun to pall, and it was hard living on a public servant's meagre pay. Charles gave Alice a dog which she called Blossom, but it couldn't make up for his three-month absences.

Alice's daughter Lorna once wrote in a letter to her nephew that Alice remained impulsive and feckless throughout her life. She was a fey, childlike figure, who never seemed to catch up with her initial impetuous decision to go to the new land. Two incidents seem to confirm Lorna's views. The first was when a twenty-one-year-old Alice was clambering up an almond tree in the orchard and fell out. She appeared unbruised, but the next day she gave birth to her second child, Charles Edward, on 7 April 1858. The idea of any woman climbing a tree when nine months pregnant seems strange; Alice was attempting it in stays and long petticoats. The second episode followed soon after. Alice was so engrossed while reading a book in the garden that she didn't notice the toddler Lizzie dragging her brother out of his cradle and on to an anthill. She didn't even look up when, black with ants, he started bawling.

After five years, the extended family, which now included a third child, Hedley Lawrence, born on 27 June 1860, finally moved to the new observatory in the West Parklands. There was room for a paddock as well as the observatory in the garden. They also had a Moreton Bay fig tree and a pepper tree for shade. The best part was the proper slate bath on the veranda and the wide front staircase. Lizzie and young Charles Edward would wrap their brother Hedley in a blanket and roll him down the stairs.

When in town, Charles was always busy with some project. He liked to compile the weather reports himself, taking the barometer readings and measurements, soil temperatures and wind directions, while pottering round his garden. As he expanded the telegraph's reach, he would add the weather reports from other Australian cities to his daily bulletin, which he posted outside his office. One story Charles often recited was about the day he was sheltering under a veranda from a sudden deluge. He saw that the man next to him had an umbrella. 'You're lucky to have that with you,' Charles remarked. 'Oh no,' replied the man. 'When the day seems uncertain I always look at the forecast and, if that man Todd says it will be fine, I take an umbrella.'

Todd had more luck with the stars, and was soon giving astronomical discourses on Friday nights. When asked why he'd become so interested in astronomy, he would reply with another of his terrible puns: 'My mother introduced me to the Milky Way.' He made a point of being the fastest telegraphist in the colony, working at thirty-five words per minute and sending up to sixty messages in an hour. While his operators often fumbled with the new Morse code, Todd could read messages just by listening to the clicking of the apparatus, rather than reading the dots and dashes. On a rare

Sunday off, Todd's children would creep into his study and the amateur scientist would let them look at his microscope, showing the intricate detail on a butterfly's wing or a mosquito's tongue.

After hours of tramping through shopping malls, we finally located the site of the observatory, only to find that it had been pulled down and replaced by a boys' school. The only thing approximating to the Todds' house was an English gentlemen's club with green leather sofas and spotted dick for pudding. The elderly doorman was unimpressed by our shorts and trainers; Aussie informality hadn't reached this outpost. Despite our pleas, we were sent away in case we interrupted the preprandial gin and tonics.

The next day was Easter Monday, and smart Adelaide was driving out to the annual Oakbank Races. Started in 1876, this picnic race meeting soon became one of the most prestigious in Australia. Charles, an avid racing fan since his days working out the odds as a child in Greenwich, used to drag Alice along. We realised this might be our chance to mingle with the cream of old Adelaide society, descendants of the Todds' friends and neighbours.

Warned to go early, we armed ourselves with tumblers so we could hop from Land Rover to Land Rover. A multitude of pink arms and legs greeted us. There were sunburnt teenage girls squeezed into pastel dresses, mothers with headscarves and ruddy cheeks, and flushed fathers guzzling champagne between the six-packs. Wellington boots, labradors and golfing umbrellas were being unloaded. A few stray helicopters were parked between the Pimms' stands. The jockeys were escorted to the start by men dressed in the red coats of English huntsmen. The girls had stocky English legs, German shoulders and sensible chins. The boys tended to ginger hair with the occasional Italian eyebrow. Everyone was of European descent. Modern Australia may now be part of the multi-ethnic Asia-Pacific Rim, but Oakbank Adelaide hasn't married it.

After the races, we drove over to see one of my relatives, Patience Fisher, a great-granddaughter of Todd's. Pat came from such old Adelaide stock that she didn't need to be seen with her picnic hamper at Oakbank and was gardening in dungarees. She owned a property in Hahndorf, in the Adelaide Hills, which she had planted with English oaks and clematis. Now in her seventies, she could remember my great-grandfather, Todd's son-in-law William Bragg, who used to take her to pick the first strawberries in the fruit cages and blew miniature glass animals for her in his workshop.

Pat had learnt most of her family stories from Great Aunt Lorna. Pat and Lorna had both suffered some terrible trauma to do with 'men'. Lorna's great love had died when she was nineteen. 'She had every knighthood in Australia after her, but she wouldn't look at another,' Pat said, and refused to divulge any more. The love of Pat's life was already married when they met and couldn't leave his wife and children. The two spinsters had become the guardians of the family history. It was Pat who had sent me Lorna's scrapbook as a wedding present.

We talked about the monarchy and the Murdochs, as Pat wandered round her musty kitchen finding chipped, pale blue mugs, scouring out the teapot and shooing the dogs off the chairs. From a stack of tins, she produced Alice's Cambridge version of Dundee cake, now flavoured with mango to take account of the local climate, and rootled around for old newspaper cuttings, photographs and family letters. She told us that Australia would have to become a republic, 'but we must fight for a president with English roots. I know the royal family are German, but we couldn't possibly have one here,' she said. 'Nor an Italian or Welshman and I'm not sure about the Scots.'

Having settled in the conservatory, with the sun flooding on to our heads, we tried to steer her on to family genealogy. She asked after my mother, another Patience, and peered at me for several minutes before confirming that I had Alice's broad forehead but little else. 'Your elder brother is meant to look like Todd,' she said. My husband also qualified in this respect, being both pale and thin and born two days before Todd, on 5 July. Having established his status as an honorary cousin, Pat was happy to flirt with him all afternoon.

My namesake didn't interest her as much as Todd, who was her hero.

'They never really recognised his talents. He had second sight, you know, and was always seeing ghosts,' she claimed confidently, before launching into several complicated tales involving her great-grandfather's psychic gifts. 'The outback is full of spirits,' she said.

I asked Pat what Todd's relationship with the Aborigines had been like. 'Oh, they got on all right,' she said. 'At first they used the telegraph insulators to sharpen their spears. But Todd soon trained them to check the wires for spiders which were a real problem, as messages kept getting side-tracked by their webs. I think Todd rather liked the blackfellows. They always referred to him as Telegraph Todd.'

Pat was a classic Adelaidian. She read the latest London books, knew the gossip from Cambridge family weddings and still made brandy butter for Christmas on the beach. But she had no interest in swapping her sun-scorched cottage and cattle for the Berkshire Downs and couldn't understand why my family had returned to chilly Britain. Taking a tour around her immaculate, box-hedged garden, it was hard to imagine what Australia had looked like when Todd and Alice first stepped off the boat. Pat suggested we took a day trip to Kangaroo Island ten miles off the coast. This, she thought, might give us an idea of Adelaide as it was when Charles and Alice first arrived, complete with koalas and other indigenous wildlife. The local settlers can trace their roots back to pirates of the eighteenth century and are proud of the fact that the countryside hasn't changed for two hundred years.

The flight took us only twenty minutes, but the difference was dramatic. Orchards, hedges and sprinklers made way for gum trees, spinifex and dust. There were seals on the beaches and wallabies in the scrub. The residents want to keep it that way for the tourists. They are terrified of anything European creeping on to their soil. If they spot a dead rabbit on the road, or worse a squashed cat, they worry that these foreign animals will ruin the island's unique ecosystem. For a joke, mainlanders sometimes smuggle a dead feline cadaver on to the island, and watch the ensuing panic. Visitors are never told about the periodic culls of the koala population, one of the few activities today's islanders still share with their ancestors.

Alice often mentioned the wildlife in her letters home, but it was only to say that she now owned some dear little sealskin slippers with bells on them and baby kangaroo gloves called 'joeys' which were as soft as kid. The longer she stayed in Adelaide, the more she wanted her home to look like Cambridgeshire. She planted packets of lupin seeds and sweet peas to stop herself feeling homesick, and persuaded an embarrassed Todd to punt her up and down the Torrens River, trailing her hands among imaginary lilies.

Food was another problem. When Alice arrived in Australia, she knew very little about cooking. According to Pat, who had become our bible on family matters, Alice's mother's cook had provided some basic instruction before she left for Adelaide, including the rudiments of making a good gravy. But her husband finally became bored with lamb cutlets every night, and employed more help. The pantry filled with strawberry preserves and cured hams, though the new cook's real expertise was jellies and creams. Todd wrote home, 'Alice and I are quite well, both getting fat.'

Back in Adelaide, we were taken to a grill house to try 'real Australian food'. My husband's starter was crocodile, emu and possum tricolore pâté, followed by chargrilled kangaroo with a lime glaze and finished off with wattle-seed ice cream. My double emu cheese soufflé was solid and felt like it had been mixed with authentic desert sand. This was bush tucker tamed and sold to jaded tourists.

My other mistake was to visit a eucalyptus factory on Kangaroo Island. We were shown how the Victorians used to boil down whole eucalyptus branches in great cauldrons. They would then add water and filter off the oil as a cure-all for everything from upset stomachs to paint removal, piles, eczema and rusty pans. The early settlers had used the remedy in large quantities as there were few doctors to prescribe anything else. Convinced that it would cure my jet-lag, I splashed the eucalyptus oil liberally into my bath. Half an hour later, I noticed ripples forming over my skin. At first I thought the water was too hot. Then I began to itch, my throat swelled, and I could hardly speak. I gasped to Ed that I was about to die of eucalyptus poisoning. The hotel doctor had to be called. I was injected with anti-histamine and smothered in calamine lotion. For hours I was scratching and vowing never to use anything again unless it came in a plastic bottle and was stuffed with artificial ingredients.

Dressing to conceal my blotches, I prepared to go out. Friends had suggested a night in the casino, not what we usually did on a Saturday night in London but apparently quite normal in Adelaide. They explained that the croupiers would be earnest students who had spent the day working in the library, and many of the games would be as innocent as watching metal horses running around a piece of felt cloth. We thought we'd go for an hour. As we entered the old converted railway station it was clear that this was where much of the city was spending the evening. Parents were at the card tables assessing their future son-in-laws. Teenage boys in tight white trousers with a girl in one hand and a drink in the other were expertly scanning the blackjack boards. Wives were massaging their husbands' shoulders while they sat at the roulette table and grandmothers were guzzling peanuts and piling chips into their handbags. Adelaide's sleepy exterior had more life beneath it than we thought. 'The holiday begins here,' said Ed. Only the most dedicated were still there when we left at five a.m.

Gambling has become the second most popular form of family entertainment in South Australia after television. Newspapers rant about the depravity of it all, but it must run in the genes. Adelaide's first settlers had gambled on Australia being a lucky country, and it was while I was helping Pat bring the horses in from her paddocks that I heard about Todd's great bet.

5

. . . . .

The Great Gamble

It took a decade soaking in the Australian sun before Todd was ready, in true Kipling style, to stake all his winnings on one turn of pitch and toss. By 1865, Todd's tentative dream of stringing together the continent by building the world's longest telegraph line from Adelaide to Darwin was becoming an obsession. He wanted to take 400 men and 40,000 poles through the centre of the continent, across an empty land, where a crowbar left in the sun for more than a couple of moments became too hot to handle.

By now Charles had established himself as an inspired astronomer, adequate weather forecaster and the most dedicated telegraph expert in Australia. His salary had increased, and the Todds' position in Adelaide society seemed secure. But he was about to risk it all. If Charles got the go-ahead for this trans-Australian project, it would be only the start of his problems. Should he fail, it would mean the end of his public service career, and he would jeopardise the lives of scores of men. He had no idea where the telegraph should go. He knew almost nothing about the territory to be crossed or the weather conditions to be faced, and he hadn't begun to think of the logistics of providing food and drink for the men and horses.

But Australia was thirsty for news. By the mid-nineteenth century, when Alice and Charles set off across the globe, the world was slowly being threaded together. After the Indian Mutiny and the Crimean War, Queen Victoria's governments were beginning to recognise the importance of instant communication. The telegraph gave them greater control over their far-flung dominions. It also provided them with immediate information of political unrest, troop movements and supply requirements.

Only Australia was not deemed worthy of a direct link with the mother-land. Letters, messages, bills and government instructions all had to come by boat via Ceylon or round the Cape of Good Hope. A regular mail service wasn't started until 1852, and the opening of the Suez Canal in 1869 barely improved matters. When Adelaide laid on a fast ship to transport the post from Albany in Western Australia straight to their port, it still only cut the time by a day. Australians knew they were unloved. As Charles pointed out to his friend McGowan in a letter he sent in 1857, 'I really think that they send us "convicts" all their refuse. For instance the telegraph wire they supply us with is made up of all their short lengths.' Private operators were equally unwilling to take the risk with the new continent.

This five-month timelag condemned the colonies at the end of the world to a second-class existence. Farmers, having sent their annual wool supply to Britain, would discover too late that the market now craved Indian cotton. Mining companies would send back opals and gold and a year later receive paltry cheques because diamonds were all the rage for hatpins and brooches on Regent Street. The new Australians had to read five-month-old newspapers. They caught boats back to Britain because a parent was dying, to be told when they docked that it had been only a cold. They were always the last to know when a new prince or princess had been born. Adelaide went into a frenzy of letter writing in the days before the monthly boat departed and the newspapers brought out special editions to send home. The new continent was desperate for a cultural, personal and economic link with the rest of the world, and Charles wanted to be the one who provided it.

Alice used to retire to her bed with a cold compress and a whole fruitcake as soon as the mail-boat had left, knowing that it could be a year before her letters were answered. She would write pages to her family in Cambridge asking about the latest bonnets and births, and the progress of her brother Edward's children. She loathed being behind on the fashion scene. Her mother, the former bonnet-maker, was one of the best dressed ladies in middle-class Cambridge. The family had saved to send Alice out with evening dresses in duck-egg-blue chiffon, lavender velvet and tobacco-brown silk. Her father had spoilt her with delicate pairs of matching satin shoes, but they soon became miserably frayed, while her mink muffler was barely touched in the heat.

The knowledge that she was so cut off from her family was the greatest strain. She continually worried about the health of her remaining family. In 1861, in desperation, she persuaded Charles to spend a substantial portion of his annual income to send her on a trip back to Britain with her four-year-old, Lizzie. Captain Bruce had offered her a berth on the *Irene*, setting sail on 19 December 1861. Five months on the high seas seemed worth it if it meant she could see her family again.

Sarah wrote a letter to Charles describing Alice's arrival. 'A few nights ago at 12 of the clock I heard the post chaise coming down the street and soon there was a loud knocking at the door. Hardly waiting to put a wrap over his night-shirt, dear Papa hurried down, while Mama and I waited anxiously on the landing. "Mama," cried Papa, "It is dear Alice and little Elizabeth!"'

Alice returned subdued. None of her fenland clan had seemed remotely interested in hearing about her promised land and she had felt left out by their gossip and cut off from their plans. Even when Alice showed them her vibrant watercolours, they didn't seem to understand. Where she had imagined a triumphant return to the bosom of her family, she discovered that her nephews and nieces were shy of her child Lizzie, her parents absorbed in their other grandchildren. In London, Alice met Todd's elder brother, Griffith, who asked her to adopt his daughter. As his wife had recently died, she felt she couldn't say no, so the seven-year-old Fanny Todd boarded the *Irene* with Lizzie and Alice. As the barque set sail, Alice knew she wouldn't return for years. When she arrived back in Adelaide, her son, Charles, by now three, refused to kiss her. Stranded between an uninterested Cambridge and an uncultured Adelaide, Alice sometimes felt destined to spend her life in the doldrums.

In Adelaide, there was pessimistic chatter about a telegraph link with London, but only Todd spent his evenings poring over possible routes, as his wife mended his breeches. While my great-great-grandparents adapted to life under a southern sun, the telegraph system had slowly spread out from Europe towards Asia, but no one thought to include Australia.

For the first few years after Todd arrived in 1855, he had been reluctant to admit his plans to anyone except Alice. After all, no one had yet managed to cross Australia, let alone string a wire from north to south. Some cartographers were still convinced that Australia's core was covered by an inland sea, other experts speculated that it was controlled by tribes of black warriors, and the romantics hoped for hidden valleys and lost civilisations. Todd had to accept that if there was to be a telegraph link it was more likely to go from the Gulf of Carpentaria in the north, down the east coast to the bigger, richer cities of Sydney or Melbourne. Adelaide, with its five main streets, would never be deemed grand enough to become the international telegram centre for Australia.

Then John McDouall Stuart, a surveyor who had made his name with Captain Sturt, crossed Australia from south to north for the first time on 24 July 1862, seven years after Charles had arrived. Stuart and his small band of men returned to a public holiday in Adelaide and the South Australian government gave him a reward of £2,000. But the trip had crippled him. His hair had turned white, he was nearly blind from the sun and he had suffered badly from scurvy. Alice was shocked when she was introduced to this wizened, wheezing, arthritic testament to the hardships of the interior. But she knew that her husband would not easily be deflected from his dream, and while she had no interest in the mechanics of the telegraph, she had begun to believe that their future happiness would depend on the success of Todd's plan.

Todd, meanwhile, was ecstatic that the continent had been crossed. He grilled Stuart about the minutiae of the trip, wanting to know exactly how wet the 'Top End', near the embryonic port of Darwin, could get, and whether there was enough water for horses in the centre. Stuart reassured him that there was no sea in the centre, and that, with only one exception at a place now called Attack Creek, the few Aborigines he had met seemed friendly. Satisfied that he could do it, Todd wrote to the Governor: 'The erection of an overland line to the north coast should be regarded as a national work, in the carrying out of which all the colonies should unite.'

In the early 1860s, the telegraph wires were creeping from Europe towards India, opening up three options for Australia. The first would be a submarine cable from Jakarta (then Batavia) to Moreton Bay on the east coast, with a landline to Sydney. The second would be a submarine link to Perth, the capital of Western Australia. Finally, there could be a submarine cable to the north coast followed by a landline to Queensland. An overland line across the centre to Adelaide wasn't even discussed.

Whichever the route, the cost would be astronomical. Lionel Gisborne, an American cable agent, thought he had the answer. With Cyrus W. Field, the telegraph entrepreneur, backing him, he had managed to raise over £1 million to lay a cable from Britain to America. He was a spiv, but an impressive one. Capitalising on his initial transatlantic success, he floated a new enterprise, the Red Sea and India Telegraph Company. Favouring the Queensland route as being likely to make him the most money, the entrepreneur wrote to Lord Stanley, Colonial Secretary in London, with his proposal.

Gisborne decided to send his brother Francis to butter up the colonials and persuade them that the scheme, which was going to cost £800,000, was a bargain. The project would be a joint private/public initiative. The colonies would contribute a proportion of the building costs and would also pay Gisborne six per cent of any revenue they collected for telegrams sent.

Charles seized his chance. He was convinced that an overland route through the centre would be only a quarter of the cost of a line along a swampy coast. Francis Gisborne had to land in Adelaide first, and as soon as he set foot in town Charles steered him to the bar of the only smart hotel and, over a rare bottle of imported claret, laid out his plans. He was confident he could turn these new Adelaidians into a relay team of 'pole-putters' across the largest expanse of arid country in the world. By constructing his telegraph wire straight across the interior, he believed he could bring Australia into contact with London in seven hours rather than five months. Gisborne went home complaining that the colonies had 'opinions of their own, a most discouraging factor'.

In an editorial on 20 April 1857 the *New York Herald* had called the laying of the telegraph around the world, 'the great work of the age'. *The Times* was even more gushing, writing on 6 August 1858, 'Since the discovery of Columbus nothing has been done in any degree comparable to the laying of the transatlantic line.' Tiffany's, the New York jewellers, even bought the left-over pieces of cable and sold them as souvenirs. But the sixties proved a decade of frustration for telegraph enthusiasts. On Gisborne's return to Britain, he was told that the transatlantic cable had snapped. In 1862, the brothers tried floating another company, the Anglo-Australian and China Telegraph Company, but investors in underwater cables had already lost £2¼ million in broken lines. It wasn't until 1866 that rubber was improved to such an extent that the Atlantic cable worked efficiently. Cyrus W. Field was praised for having 'moored the new world close alongside the old', and there were simultaneous celebrations in London and New York. The opening of the transatlantic route was followed by an underwater cable from Suez to Bombay. India, the brightest jewel in Britain's Imperial crown, had been connected.

It slowly dawned on South Australian politicians that a telegraph system across the centre would not only unite them with England, but also with the new territory that the southern colony had claimed in the mangrove swamps on the northern coast. The line would be the perfect link between the north and south and would cement the future of the fledgling Northern Territory. Todd's plans began to be taken seriously.

Bolder and brasher Queensland was adamant that the line should be theirs. But the colonies had to wait another two years before anyone thought about their plight again. Then, in 1869, the Telegraph Construction and Maintenance Company, which had laid the line to India, decided to try stretching it to Australia. Captain Sherard Osborne, the managing director, favoured the Queensland route. In London, MPs were dismissive of a telegraph link. In Adelaide, some politicians developed cold feet, with one MP calling it an 'egregious piece of insanity' and another a 'blatant white elephant'. But the new Governor of South Australia, Sir James Fergusson, didn't want a British–Australian link that bypassed his colony.

The Times took up the idea. On 9 December 1869, it published a letter from a man named Dalrymple Ross, arguing that the size of trade with the Australian colonies necessitated a link with the home country. Ross pointed out that by 1867 the population of Australia had grown to over one and a half million and the number of livestock to fifty million. The island had two million acres of land under cultivation. The total number of letters sent the year before had been nearly twenty-seven million, and the mining interests had already made £5 million. 'It must be remembered that the colonies have no Rothschilds or Marquises of Westminster – that their wealth is of this century, that it is not inherited nor accumulated wealth – but it merely represents the enterprise and indomitable perseverance of a single generation,' Ross wrote. The editor of *The Times* responded with a leader saying a line should be established immediately and politicians began to look interested. The only problem was that the project didn't include Adelaide as the hub of the new system. Todd, sitting on the other side of the world, must have thought his wedding-day dream was at an end. The South Australian parliament, realising it was about to be sidelined, sent a forlorn letter to Osborne's company saying the colony would be prepared to pick up the tab for a landline itself.

The letter missed Noël Osborne, another brother sent to woo the colonies. Luckily, he also landed in Adelaide first, on 11 April 1870. Public interest was now so great that the newspapers devoted pages to his arrival. Todd tried his claret trick again. He told Osborne that crossing the centre would be cheaper, contending that South Australia only had to deal with desert while Queensland had to negotiate jungle. Todd was so desperate to get the deal that he promised to deliver the landline in two years for only £120,000, provided the cable was brought in at Port Darwin. The South Australian government backed this up with a letter. The company's representative held his tongue until he had talked to Victoria and Queensland, but he knew he had been offered a bargain.

Queensland was furious. William Cracknell, their superintendent of telegraphs and brother of Charles's assistant, wrote to the British government saying that the South Australians were deluded, and in any case his state had already built some of their landlink.

Charles became insufferable, refusing to join meals at home, and Alice could barely cope with the tension. By now she had given birth to another child, in 1865, Alice Maude Mary, and with four children under ten she needed Todd's support. But she was also increasingly desperate for a link to her homeland. When her father died in 1865, he didn't even know that he had a new four-month-old grandchild. Alice received a copy of his funeral hymns six months after the event. The next year, she received a letter saying that her brother Alfred had died, and she hadn't even realised that he had been ill. Sarah tried to keep her sister's spirits up with lengthy letters about the nephews and nieces, but in the winter of 1867, she herself caught influenza, and died of complications that May. Alice was distraught when another black-rimmed letter arrived in October. If only she had known Sarah was ill, she would have taken a boat home. From a family of eleven, Alice now had only one brother left. She needed the telegraph to stay in touch with her homeland, and check that her remaining family were still well.

The Osbornes were in no such hurry. They were happy to play the two colonies off. On 4 June 1870, Francis wrote to the latest South Australian government: 'I am now in a position to state that the cable will be landed at Port Darwin, if the South Australian Government will pledge themselves to have a land-line open for traffic by 1st January, 1872.' But he was still in negotiations with Queensland, telling them that the company would lay the line to Normanton if Queensland would guarantee five per cent of the cost of the additional cable. The Queensland government also agreed.

The South Australian government passed the relevant bill with a large majority. Only a few dissenting voices muttered that this was a 'childish game of opening your mouth and shutting your eyes'. Todd was told to press ahead. By now he had also been made Postmaster-General, in charge of all the mail in South Australia, so he had his work cut out. The continent hadn't been crossed since Stuart's expedition seven years before, but Adelaide soon rallied to the cause. The Chief of Police was asked to choose the sturdiest horses, and held a competition to see how much sand they could drag up a dry riverbed. Charles ordered a hundred Afghan camels with their drivers from Egypt. A saddler made 27 sets of heavy harnesses, 45 pack saddles, 300 collars, and many map cases, instrument cases and haversacks for the officers. He used over six hundred hides and employed forty extra men.

The line would consist of a single strand of No. 8 gauge galvanised iron wire purchased from Johnson and Nephew in Manchester. They would also need insulators from Germany, and British batteries and relays. In ten weeks, one company at Port Adelaide made 30,000 insulator pins from ironbark, a tough Australian eucalyptus. The bakers worked round the clock to produce hard-tack biscuits. And a farmer from Booyoolee Station invented a way of preserving beef chunks in gravy which they canned with red labels and nicknamed bully beef, so creating a staple which served Australia and the British Empire for a hundred years, although the French have always disputed this, insisting that the word came from "bouilli" – the boiled beef made for Napoleon's army.

Trollope wrote of the extraordinary scheme in his book on Australia:

The work had to be done through a country unknown, without water, into which every article needed by the men had to be carried over deserts, across unbridged rivers, through unexplored forests, amidst hostile tribes of savages, in one of the hottest regions of the world. If the gentle reader will only think of the amount of wire required for 2,000 miles of communication, and of the circumstances of its carriage, he will, I think, recognise the magnitude of the enterprise.

Despite all South Australia's efforts, it looked as though Queensland was going to win the contract. Their line had to be only a third of the distance and their treasury was flush with cash from cattle and gold. South Australia had almost nothing in its coffers and was being drained of good men by the day, as they went off to pan for gold in other colonies.

Todd soldiered on. The company kept playing off the colonies against each other until Adelaide promised that if it went over the eighteen-month deadline it would start paying penalties – £70 for every day's delay. The contract was signed in June 1870; the line had to be open on 1 January 1872. The town of only 184,000 would be ruined if anything went wrong. Disraeli had already said that the colonies should be cast aside, being a 'millstone round our necks'. They should be made to sink or swim, as 'they always take and never give'. The Little Englanders in Westminster would be unlikely to bale out this barren land.

Todd wrote in his diary: 'Then, perhaps for the first time, I fully realised the vastness of the undertaking I had pledged myself to carry out . . . It was my life's ambition which I had eagerly looked forward to, but now that its weight really rested upon me, I must confess, at times, it seemed too heavy to bear.' South Australia's future was suspended on a thin metal wire.

6
— · · · ·

Ditching the Camels

Ed never got to sunbathe on a beach in South Australia. We spent the entire holiday pursuing the Todds, and my ancestor had seen quite enough sand in the outback to keep him away from the beach. In fact, in Charles and Alice's day there had been an inspector of nuisances who ensured that men and women were allowed to bathe only if swaddled head to toe in shapeless tents.

The three weeks we'd spent in South Australia were leisurely in comparison with most of my assignments as a journalist, but I'd become obsessed by returning to finish my great-great-grandparents' story. Charles and Alice were part of the Victorian crusade to extend British civilisation and propagate civic virtue. My working life, in comparison, had been lived at one remove. In Bosnia or Beirut, I was always following someone else's adventure or misadventure. As an interviewer, I wrote about other people's families, and as a political reporter I was either in Middle England's shopping precincts finding out which way people would vote, or in the Palace of Westminster watching others try to run the country.

I suggested to Ed that we should take a sabbatical and follow Charles's exact route across Australia, on horseback or camels. It couldn't be that hard to retrace Todd's expedition. Ed immediately ruled out any animals, insisting they would give him hayfever and saddle sores. He was more tempted by the idea of taking a four-wheel-drive through the outback. This was the closest he was going to get to his American-road-movie dream of driving an open-top chrome-bumpered Caddy across the USA. He claimed he knew how to change a tyre and fix a fan belt, whereas I couldn't shoe a horse. There was nothing wrong with today's Lawrence of Arabia looming out of the heat haze in a Toyota, said Ed. We could load up with water and bully beef and head out into the scrub.

I finally ditched the camels. We could take time off only over Christmas. The problem was that this would be mid-summer in Australia. The temperature would rise to over forty degrees, the flies would be hatching and the mosquitoes' blood would be up. But at least we weren't expected to plant any poles. From the vantage point of London's Notting Hill, it seemed perfectly straightforward.

I wrote to our relatives. There was no reply. Then I received a terse letter from Pat, containing a newspaper cutting about our proposal under the title 'Mad dogs and Englishmen'. Our simple plan was evidently so ludicrous that it had reached the gossip columns of the *Advertiser.* 'Best bring a hat, Alice,' it warned. Pat was even more severe. She told me that tourists were always getting themselves lost in this kind of heat, gallivanting around off the beaten track. At least one party a year died of heatstroke, many others had to be rescued by air. She didn't want us embarrassing the family. Couldn't we just head straight up the tarmacked Stuart Highway and cross the continent that way?

But this metalled strip went over a hundred miles to the west of the route that Charles had followed. If we were going to use Todd's original maps, we had to head straight for the centre through the desert. Ed, having won the battle of the four-wheel-drive, saw the opportunity for another high-tech gadget, a G.P.S. (Global Positioning System). 'Todd would have loved this,' he said when I protested that we were supposed to be emulating the Victorian explorers. This instrument, he explained, would bounce signals off a satellite to determine precisely the car's longitude and latitude. Taxi drivers in London use it to pick their route around traffic jams. Round-the-world yachtsmen plot their course with it. The G.P.S. is the offspring of the compass, Ed rationalised. I drew the line at a satellite telephone.

Seven weeks before Christmas, I flew ahead to Adelaide to make the preparations. As well as the G.P.S., I brought detailed maps, penknives, torches and factor 30 sun-cream. I had no idea what we would find along the route.

The airline lost my suitcase somewhere between London, San Francisco and Melbourne. Four days later, they gave me an almost identical one, but it was stuffed with brand new baby clothes. I rang to complain. First Shirley, then Kerry, then Debbie, Diana and finally Keith, told me I was being difficult. The baby clothes were definitely mine. I could do what I wanted with them. So that was the last I saw of our G.P.S.

I was starting from scratch. Once Pat realised I was serious, she helped out. Dresses and trainers, she said, were fine in London, but in the outback I needed jeans, long-sleeved shirts, Chelsea boots with tabs on the back, and an unadorned Akubra hat – 'the corks are only for tourists'. She'd once spent a month neutering lambs and branding bullocks on a property in New South Wales, and knew everything. Tents would be too hot, we should buy swags – canvas sleeping bags peculiar to Australia. She found some disposable paper knickers in a hardware store which she thought were a wonderful idea. Of more use was the white Toyota she procured. She pinned a large red tea-cloth on the aerial so we could be seen by other travellers as we crested the sand-dunes.

Jane, our Adelaide contact, had now left the tourist board, but made out lists of places to stay when we had had enough of our swags-in-the-dust lifestyle. The guide shop provided us with modern maps, drawn up by satellite rather than sweaty surveyors, proving that there was indeed only one highway across the centre – which we wouldn't be taking – and a few dust-tracks between properties. The assistant threw in a small book on such

SAS survival techniques as how to skin a snake, before wishing me luck.

When Pat saw the front page of the *Advertiser* featuring a man being rescued from his crashed plane after two days in the centre's salt-pan lakes, she started worrying again. 'It cost A$200,000 to rescue him, I'm not spending my taxes on you two idiots,' she insisted. I promised we'd take forty gallons of water and bought an old two-way radio. Pat tried to teach me the etiquette of the airways. Sitting in the kitchen, she boomed down the mouthpiece to summon an imaginary flying doctor service. But all this Roger, Roger stuff – she had trained in the 1940s – merely conjured up images of Pat in an ill-fitting army uniform. In desperation, she suggested that we learn Morse code; and so I dotted and dashed my way back to my hotel.

The next stop was the State Library of South Australia. The main library, the Mortlock, was oak-panelled, with cricketing memorabilia from the Edwardian era and photographs of Adelaide's changing thoroughfares. In the past few years, local families had been urged to bring forward any family mementoes to provide a social history of the city. Diaries from the first hairdressers, tram-conductors and funeral directors were all deposited here, as well as letters from Todd to his wife when he was in the bush, and the journals of some of the men who had set out with him on the overland project. Having rung ahead to warn them, I found a neat pile of papers wrapped in pink binding with a note on the top saying, 'For Alice's granddaughter'. In another room I was shown a filing cabinet filled with spidery notes made by Major-General Symes, who died before completing his biography of Todd.

Local historians invited me round for chicken-balls, pork rissoles and wine sweetened with orange squash. Softened up by the drink, I listened to them explaining their pet projects, but they knew little about the Todds. One lent me his copy of a four-wheel-drive magazine, and warned me never to leave the car if I broke down – that way we might at least be spotted from the air before we died of thirst. Most Australians, he explained, clung to the coast and wouldn't understand why we wanted to get lost in the interior for the sake of finding a few ant-eaten poles. Another, who had a mesmerising habit of picking his dandruff and eating it, kept shouting at me every time I said farm. 'In Australia it's a property, for God's sake. Not some airy fairy thing with chickens and ducks.'

Next I went to Adelaide's History Trust of South Australia, which still held some of the old Telecom (now Telstra) archives. The place was in chaos, as the trust had decided to relocate to a new premises. As I watched, boxes were being wheeled away with remnants of Todd's life inside. 'You can apply to see the things when they've sorted it out in a few months,' insisted the removal man. But that would be too late.

Slipping into the lift, I descended to the cellar and started rummaging. In an alcove I found Todd's bookcase, a few astrological texts and an old prayerbook with the inscription, 'To my dear Alice, from her admirer Charles Todd'. So that was one legend made fact. In a three-inch leather notebook I also found his shopping list for the journey to Australia. Charles had not only written down Eliza's hairpins and Alice's toothbrush, but also bedding, crockery and a necktie for himself. In another corner was his post-office desk. I found his instructions to the overseers mixed in with a batch of 1960s guides to Alice Springs. Pat thought the mayor of Adelaide might have Alice's music book, the Freemasons had his compass and field glasses, and a cousin had his napkin ring. 'I've got his riding whip,' Pat said. 'The museum tried to yank it off me but I use it round the property.' Barry Todd, another great-grandchild, was thought to have a copy of Todd's will.

The local radio station invited me on to its morning programme to see if I could track down some of the relatives of others foolish enough to join Todd's scheme. After my introduction, the phones started ringing. Everyone in Adelaide seemed to have been related to someone who helped in the operation. The programme presenter tried to marshal the callers into areas: food-providers, officers and camel-men. There were also callers such as the woman who thought her great-grandmother had been the governess for Alice's children and reminisced about how they got her piano to the Springs on the back of a camel. I was convinced that Alice had never been to Alice, but the woman was adamant. She now had the piano, and it took pride of place in her sitting room. Another man collected porcelain insulators from the line. He had one with Todd's signature, but most were 'just the bog standard variety; they look pretty in the rockery'. Another family had Todd's chair, which he had given them in return for their portable telescope.

Between them, they painted a picture of an extraordinary enterprise. Todd wanted the whole operation to remain firmly under his control. But he was told to split the project into three sections just under seven hundred miles long. The northern and southern sections were put out to tender and supervised by a government overseer, while the central section remained under Charles's command. The eighty men chosen for the top section were immediately shepherded on to ships for Darwin. The central section was divided into five sub-sections, A to E, each with its own overseer, and a party was to be sent out in advance to map out a route.

The southern section had the easiest job. They just had to make it up to Port Augusta and start digging. With the men assembled, Charles told them he had every confidence in their energy, good conduct and their ability to carry out the work in the time they had available. The plan was to follow Stuart's route up to the Roper River in the north, and then branch out to Port Darwin through uncharted territory. Stuart's journals weren't much use. He had drawn his maps when nearly blind, and his small group had followed a narrow, meandering route.

Todd's teams would have to cut a broad, straight path for their poles. They would be mauled by crocodiles, bitten by snakes, addled by the sun and harried by natives, all the appropriate challenges for Victorian explorers. When the advance team got to the centre, they were living on a small marsupial called a bandicoot. 'They must have been mad to try it,' I told the radio presenter. 'You'll be lucky to make it half the way across if you don't know your sump from your starter motor,' he replied.

I had Ed, who knew all about cars. Unfortunately, he had brought the flu over from England when he arrived two days before we set off, and spent the first forty-eight hours sweating under the hotel sheets. The doctor, when summoned, proved more interested in my ancestor than Ed's illness, and after a cursory glance down his patient's throat, admitted that he'd gambled his savings on a new gold mine along our route.

Finally rousing himself, Ed sneezed and shivered in the forty-degree sun, barely making it to the supermarket to haggle over the supplies for the first leg. My unfair reputation for dumping boyfriends at the first sign of physical weakness, particularly on holiday, worried Ed. He reminded me about a predecessor who had foolishly developed a cold on a skiing holiday, a terminal error. I insisted that it was only because the boyfriend had whinged so much about a snivel that he had had to go. Ed, I knew, was made of sterner stuff.

Choosing our supplies took hours as we wound up and down the aisles. Ed slouched over the trolley, trying to decide whether he could face muesli with soya milk for breakfast. Our options were limited by our lack of a fridge. Between bouts of coughing, Ed insisted we had to fill up the cool box with several bottles of chilled local Chardonnay and he needed some whisky to go with his cigars. As we had only a small gas stove, suppers would be limited to tinned tuna and pasta. We'd probably end up eating dried fruit and nuts; perhaps we should have asked for the recipes for the desert soufflé and griddled crocodile steaks at the grill house. With Ed curled up on the back seat, now half delirious and more concerned with his health than his marriage, we set off up the highway, following the route taken by the southern and central teams. The first group started up the coast, conveniently close to the German wine country.

When the men for the central section set out on 29 August 1870, Lady Edith Fergusson, the new Governor's wife, presented them with a supply of reading books. Crowds cheered them on their way. Their instructions were that they could have one glass of beer to help them pick up the pace, but the convoy wouldn't wait for them. The overseer for each small band of men was always a teetotaller who could keep discipline.

We had no such rules. The Barossa was said to have some of the best

wine in Australia, and I thought a vineyard crawl might perk Ed up. But all
he wanted was ice cream for his throat. We headed past a sign of an
accordionist in his feathered cap, and into the nearest town, Lyndoch.

The Barossa is a thigh-slapping Bavarian theme park gone mad. Streets
were called Nietzsche Drive rather than Shakespeare Road. Middle-aged
women served lunch, their plaits swinging dangerously over the frankfurters
and sauerkraut. Jolly butchers waved *leberwurst* and *lachschenken* in our
faces, and the funeral parlours had names like Brandt & Brandt. Having
spent my childhood holidaying in Switzerland, wearing lederhosen and
looking for cowbells, I rather enjoyed a little Mittel-European tradition. Ed
was indifferent.

We drove on to Tanunda, home of four Lutheran churches. The local
museum, housed in the old telegraph office, had a miniature model of a
Wagnerian castle, and a map detailing the German origins of the inhabitants. But this Germany looked menacing and swollen. It seemed to include
large parts of Poland, Czechoslovakia and Alsace-Lorraine. The incon-
venient realignment of frontiers post-1945 had been ignored. We asked the
attendant what had happened to the Barossa during the two world wars. 'It
was very quiet,' she said, leaving us to discover that many of the locals were
packed off to internment camps and the village names were changed. With
Australian soldiers dying abroad, the politicians decided that driving down
Bismarkstrasse was inappropriate.

We wanted to see the Barossa's birthplace, so we drove on to Bethany,
where twenty-eight Lutheran families arrived in 1842 and partitioned the
land in the same way as in Silesia, in long strips from the track to the river.
Our chalet had fluffy pink duvets. Toadstool chocolates melted on the
pillows, and the pond outside our window had two plastic swans floating on
it. If parts of central Adelaide were clinging on to their Victorian past, the
Barossa appeared determined to hang on to its nutty King Ludwig.

The next day, we made our way up to the Clare Valley, otherwise
known as Mid-North South Australia, to stay with Greg and Bill, friends of
Jane, who lived next to a nineteenth-century Jesuit church and vineyard.

'All those vagina pink houses, how did you cope?' said Greg when he heard we'd been staying in the Barossa. Our new hosts had set out to make the perfect old-fashioned English country hotel in the middle of a dustbowl. The early-twentieth-century beamed house was covered in chintz roses, pot-pourri and wrought-iron bedsteads. The kitchen had an oak dresser and a Mrs Beeton cookbook.

Greg was dressed in purple and had just had a blazing row with Bill, who had left in a huff but kept phoning. He offered us shortbread, gave Ed far too much sympathy, and introduced a Japanese couple from Sydney. Dressed in matching red Argyll socks and yellow shorts, they seemed unnervingly laid-back for Tokyo expatriates. We discussed whaling, the ethics of eating tuna fish, and tattoos as an art form.

Greg cooked us a four-course Pacific-Rim meal, starting with chilli squid with whitebait, followed by roast squab with fennel ravioli in truffle-scented sauce, and finishing with hot kumquat soufflé and vanilla-bean ice cream. Ed drank some home-made chicken soup. After a few tastings of Clare Valley wine, the floral print wallpaper became a blur. The Japanese began extolling the virtues of Australia and said it was such a relief to find a country with so little ideology and no glorious or embarrassing wars. Japan, the man said, carried too much baggage. His wife felt the same. She liked the gyms and sipping coffee by Sydney Harbour. But the Japanese man dismissed the Australians' ambition to become the new financial melting pot of the Far East. He was more interested in their putting greens, and insisted that Australia's future was as an upmarket retirement home.

Greg was appalled. His parents had come from East Grinstead in the 1960s with their three sons. They thought Australia would make a perfect nursery and were hoping to turn them into surfing, cricket-playing Aussies who would marry wholesome sheilas. Instead, they had all succeeded by playing up their English roots. One made old quilts, Greg ran the hotel, and the last had an antiques business. The idea of retiring in brash Australia amazed Greg. He wanted a beach hut in Kent, like his gay British hero, Derek Jarman.

We went for a walk over the hill and were confronted with miles of flat plain. The earth in the vineyards was Devon red, but as soon as the sprinklers stopped it took on the texture of solid rock. The telegraph men must have gazed on this landscape, stretching out over 2,000 miles, and wondered whether they would ever get to the sea at the other end.

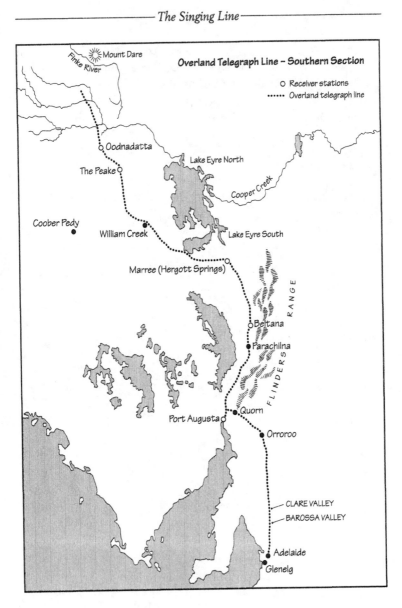

Overland Telegraph Line – Southern Section

○ Receiver stations
...... Overland telegraph line

Mount Dare
Finke River
Oodnadatta
The Peake
Lake Eyre North
Cooper Creek
Coober Pedy
William Creek
Lake Eyre South
Marree (Hergott Springs)
Beltana
Parachilna
FLINDERS RANGE
Port Augusta
Quorn
Ororoo
CLARE VALLEY
BAROSSA VALLEY
Adelaide
Glenelg

7

— — · · ·

No Commanding Personality

'He had no commanding personality,' my great-grandfather Sir William Bragg wrote when asked to contribute to Todd's entry for the *Encyclopaedia Britannica* after his death. 'No commanding personality' became our favourite insult when we couldn't agree on where to stop for the night.

It was true that Todd didn't have the Sandhurst self-confidence of an imperial explorer like Younghusband or the public-school assertiveness of a Scott of the Antarctic. What he had was the energy and determination of an optimistic age and the Victorian belief in persistence. He wasn't taking up Kipling's burden like other explorers or playing any great game with Russia. His was an exercise in civic virtue and scientific endeavour. It wouldn't bring credit to the Empire, but it could turn Adelaide from a backwater into a thriving metropolis. As his son-in-law Sir William Bragg pointed out:

> At first glance it might have been difficult to discover the source of his power . . . but those who worked for him soon recognised his sense of proportion, his strong grasp of essentials, his acute understanding, and untiring energy . . . The whole of his department was infected with his sense of duty and loyalty, his kindly courtesy and good humour.

Todd and his men must have needed every ounce of determination as the teams headed north from the Clare into the arid country beyond. As we drove out of the valley, the remaining vineyards soon gave way to yellow prairies swaying in the heat. Large, round hay-bales were the only upright landmarks. Trees, hedges and fences had vanished. After two hours, we came across a small gathering of trucks at the side of the road where they were selling sheep and beer. The men offered us a can, but this wasn't some picturesque rodeo, and they didn't want us hanging around while they haggled over the price of ewes.

So we drove to Burra, once the pre-eminent mining town in South Australia, attracting miners from as far afield as Cornwall. The café still sold pasties and Cornish cream teas, but only if you were prepared to listen to a guided tour starting with the story of a shepherd picking up a piece of ore and ending with his successors dying of tuberculosis while the British owners cashed in. Australia may not have a long history, but what they have is often political.

It was too hot to take sides. So we drove on, parched, our arms already burnt from hanging out of the window, and arrived at Orroroo at the same time as the schoolbus. In the nineteenth century Orroroo prided itself on being on the right side of the outback. The boulevard still has aspirations to being a grand thoroughfare, with a few well-watered marigolds clinging to the edge. A group of women in white hats, flat white shoes and white stockings were sitting on the benches. They looked like a convention of chiropodists, but explained that they were the local bowls team. We queued up for ice creams alongside strapping pinafored schoolgirls with names like Cynthia and Cassandra. They carried new tennis rackets and their mothers had just been to air-conditioned aerobics classes. These wheat-belt property men are doing well, but this will always be a small town.

Orroroo was Todd's favourite stopover on his visits to Port Augusta, if only because it allowed him to crack the same pun each time he visited: 'Why should a two letter town need a post office?' he would say to the innkeeper. We tried out the joke on the postmistress, who shooed us away.

Finding a camping spot before the light faded proved tricky. Orroroo is encircled by wheat. The only grass was clinging to people's backyards and closely guarded by dogs with dingo ancestors. Discovering a lay-by, we settled down to cook our supper, then realised that lighting a stove in this tinder strip would be unneighbourly. The air was still hot and harsh. After returning to town for more ice cream, we lay on top of our swags and by torchlight read Todd's 5,000-word instruction manual to the workmen.

Rationing was his first heading. How does one set about provisioning a large body of men to work for eighteen months in unknown country? Todd thought it out carefully, convinced that illness on the line would be his greatest enemy.

Each man's weekly rations included:

1lb of flour

1lb of biscuits

8lb of meat

2lb of sugar

¼lb of tea

½lb of peas or oatmeal

½lb of rice or pearl barley

1 gill of vinegar

½ gill of lime juice

2oz salt

½oz mustard

½oz pepper.

A few months later, back in London, I was sent to a health spa for a weekend to write an article. Brandishing the list of supplies, I walked into the dietician's room in my fluffy white dressing gown.

'Does this constitute a balanced diet?' I asked.

'You haven't got enough dairy products here. There's no milk or yoghurt, so you aren't getting enough calcium or vitamin A. You need some fresh fruit and vegetables to up your vitamin C intake, but there's a lot of iron, protein and vitamin B complexes,' she told me. 'Are you an athlete?'

I explained this wasn't my diet, but a nineteenth-century pole-putter's.

'You'd have been consuming about 3,800 calories a day, which sounds all right for a manual worker, but far too much for a desk job. Considering the circumstances, this is a pretty balanced diet. The risk of scurvy would be minimised by the lime juice and they'd be replacing all that salt lost in sweat. We should try the exercise regime out on some of our clients, get them building a few fences for the local farmers.'

Todd was proud of the provisions, which were managed by Harley Bacon, the son of Lady Charlotte Bacon, the Ianthe of Byron's poem. As a young girl, Charlotte Harley, the daughter of the Sixth Earl of Oxford, had met Byron, who had been struck by her extraordinary beauty. Having later married a major-general who had fought at Waterloo, Lady Bacon settled with him in Adelaide, where she became the grande dame of the social circuit. Bacon, an amiable man whose position owed more to his mother's standing than his organisational talents, was assisted by William Blood. Each sub-section also had a storekeeper and men to look after the livestock. As well as food, rations included ½lb of soap and ¼lb of tobacco each week, and a pipe and box of matches every fortnight. Bacon and Blood drove 2,000 sheep up the line to supplement the bully beef. One government official, on being shown the storerooms, complained, 'I would strike my pen through half of all these things. An axe and a quart-pot is all these men require.'

The next instruction was to the overseers, telling them to make sure that any man in charge of firearms kept them properly oiled, and giving a detailed account of how to load and fire a Colt or Whitney revolver. Before he came to South Australia, Todd had only fired unsuccessfully at albatross, but ten years of inspecting lines and shooting birds off the wires had made him one of the best marksmen in Adelaide.

The overseer for each section was given the power to suspend any of the officers or men in his party in case of gross disobedience or incapacity. He was also expected to keep a diary giving a detailed account of their movements and conditions. The sub-overseer had to keep a timesheet showing the occupation of men at all times, the number of poles cut and the length of wire erected each week. Individuals had responsibility for their tools and would be charged directly for their loss.

'Health and Morals' was the next title. 'The overseer should especially direct that anyone suffering from diarrhoea should report himself at once, so that proper means may be taken to stop it. This complaint, if neglected, is apt to run to dysentery and cholera, as well as being one of the symptoms of camp fever,' Todd said. Gambling, profane language, swearing and quarrelling had to be stamped out, and overseers had to endeavour to promote rational amusements among the party. There was a whole section marked 'Men Not To Be Kept Idle'; they were allowed to rest only on Sundays to attend a religious service.

Todd told the overseers that they must make detailed reports of any accident, signed by witnesses. The men couldn't sue for the loss of a limb. If they were killed, their widows knew they would be well provided for, but Todd realised that too many accidents would sway public opinion back home. Each party had a muster roll at dawn and again at dusk and no one was allowed to leave the area without the overseer's permission. No extra expenses were allowed on the road, so even if the men did find a shop in a mirage, they couldn't have afforded a beer.

Carts needed to be balanced before starting. The wheels had to be greased, the harness polished and the horses properly shod. Any dried dirt and sweat had to be brushed off the riding horses each evening and the stockmen should check for sores. 'No horse is on any account to be struck on the head or legs, nor punished severely except for positive vice,' Todd wrote. 'Teamsters must remember that more is done by kindness and by humouring the horses than by blows, and that the safety of the party may depend upon keeping the horses in good working order.'

Todd must have been working secretly on these rules for years. There were instructions for watching the stock at night and on the best way to tether them. 'In a country frequented by blacks, every party must keep a watch at night,' he wrote. 'It is a good plan for the cook to keep a morning watch, as he can light the fire, and prepare breakfast during his watch, and thus expedite the starting of the party.'

Using his mathematical skills, he drew elaborate sketches for constructing bridges and traversing creeks. 'Parties to travel in cool of dawn and dusk or by moonlight' would be taken for granted by all but the heat-dazed. Only Ed and I, worried about hitting a kangaroo in the twilight, were planning to drive in the midday sun.

Drawing on Stuart's experiences, Todd gave detailed accounts of how to find water. 'All bright green places should be visited,' seems obvious even to a Londoner, but other tips would be valuable if we were stranded. 'Reeds are almost always a good sign of permanent water. Rushes usually only indicate where water may be found in the wet season.' We would need to work out the difference.

Some of Todd's advice was repeated in our SAS manual. 'A person who loses himself seldom follows a straight course,' they both suggest, and recommend lighting fires instead of flares to pinpoint a camp's whereabouts if someone gets lost. Todd added that any tracking party should draw arrows in the ground pointing in the direction of the camp, so that if the lost person stumbled across them he would know which way to go. 'Cunningham, the naturalist on one of Sir Thos. Mitchell's expeditions, would probably have been saved had some such precaution been taken,' he warned. Todd was convinced that one of the reasons most expeditions into the interior had floundered in the 1850s and 1860s was because the explorers had welcomed painters, scientists and naturalists in their parties, who slowed them down.

The Superintendent of Telegraphs didn't need to be physically brave. Heroics could be left to the advance troops. What was required was the kind of nit-picking attention to detail beloved of tax inspectors and Swiss mountain guides. Every insulator pin had to be accounted for and every saddle bag counted. Todd was meticulous and methodical. 'Need I tell you how many sleepless nights and anxious hours I spent as all the apparently insoluble difficulties stared me in the face,' he wrote. But as the trip progressed, many of Todd's carefully crafted rules were discarded along with the excess baggage.

The southern section started poling at Port Augusta on the Spencer Gulf, which we had been warned by Adelaidians was 'just like Glasgow'. It was expected to meet the central section near Oodnadatta. The northern teams, already heading by boat for Darwin, would meet the central section near latitude 18.

We expected to drive through tower blocks to reach the port, yet circling the one-way system we couldn't find any grim housing estates; only pastel bungalows, a playground in primary colours, a yachting club and a croquet lawn. This was supposed to be the state's seediest port, but it was charming. There were a few tankers in the distance, but otherwise this southern hub looked like a Sussex coastal town, with zimmer-frames on the zebra crossings and matrons bowling on the green. Even the shopping mall had a man scraping chewing gum from its pavements.

Edward Meade Bagot won the contract for the southern section. His line would start in country inhabited by a smattering of pastoralists, before squeezing between the dry Lake Torrens and the wooded hills of the Flinders Ranges. But he also had to circumnavigate Lake Eyre, a salt pan so large and bleak that it was used for Donald Campbell's successful attempt to set a land-speed record in 1964. Bagot's thin metal wire would have to stretch five hundred miles to the Peake, a natural spring that had been selected as a principal staging post on the road north. His incentive was the £41 he would receive per mile of line. The men set up a camp at Stirling, five miles from Port Augusta, and on Saturday, 1 October 1870, they planted their first pole, celebrating with a glass of beer.

The teams all worked in the same way. On arriving at their starting point, the overseer would ascertain the latitude and then fix the first camp some ten miles down the line and as near as possible to feeding ground and timber. The men were expected to start felling trees and planting poles immediately, while the overseer went ahead to determine a rough track to the next working party. Whatever the terrain, a path fifteen feet wide needed to be cleared for the line. All scrub, undergrowth and overhanging trees had to be cut down. Advance parties were expected to lay out pegs to mark the position of poles and the surveyor to plot them on a map. One sub-overseer wrote, 'begin work at 5.30 a.m., breakfast at 7.30 a.m. to 8.00. Leave off work at 11, begin again at 2.30 p.m., and work til 6.'

Success hung on the poles. They should be 'straight rough gum, pine, or stringy-bark saplings, or other timber not liable to be attacked by white ants, perfectly sound, stripped of the bark, eighteen to twenty feet long, nine to ten inches in diameter at the butt and five or six at the top, or square poles of the same length'. Foresters working in a Scottish pine plantation would have difficulties meeting these requirements. The men were only allowed to use 'stunted timber' if they took two 'runts', bolted them together and bound them with three bands of hoop-iron in five turns. A six-or seven-foot length of gas piping could then be cemented into the butt to increase the height. The men had to find twenty poles for each mile, and dig to four feet to secure each one. A hole had to be bored vertically into the top and the insulator pin inserted. Only then could the insulator be secured using a piece of leather. This small porcelain bell jar was crucial to stop the message leaching away down the posts. Every second pole had to be fitted with a lightning conductor. Finally the poles would be hauled into place and a team would thread the wire. Forty thousand poles needed to be erected across Australia in this way. When wood ran out, Todd eventually resorted to ordering a batch of metal 'Oppenheimer' poles from London.

How many original wooden poles were still standing, how many insulators would we find in the bush? Would we stumble across some bent pickaxe? I started dreaming about poles. Reading through the Lonely Planet guide to Australia, we came across 'Souvenirs To Take Home: Often the best souvenirs are the ones which have special meaning . . . the antique "glass insulators" picked up along the old Telegraph line.' I was incensed. This was grave robbery. How could a reputable guide be exhorting tourists to steal a nation's inheritance? Ed half listened as I ranted about the outback's Elgin marbles.

We set off parallel to Bagot's line, west of the Flinders Ranges. As the tarmacked road flattened out into plains of spinifex, the towns shrivelled up while the wheat and the cattle were lost in the dust. Despite passing the odd road sign and station wagon, civilisation was receding into the heat haze behind us.

The section leaders had no problem finding four hundred volunteers prepared to don blue shirts and moleskins and set out for the outback for eighteen months. Police had to be kept outside the doors of the GPO to ensure that those waiting to sign up formed orderly queues. Many of the men were respectable, often religious, first generation Australians who saw this as their chance to grab more land than a poky backyard. Others, like Harley Bacon, wanted adventure. He had asked his mother, Lady Bacon, to write a letter to the Governor, requesting a place. There were bank clerks, shopkeepers and cattle drovers, all desperate to see the colony's success assured and to return as heroes. But there were few labourers who knew how to wield an axe. They had to learn to become teamsters, wire-men, blacksmiths, surveyors and cooks. The pay was minimal, the conditions would be arduous and most hadn't even thought about the monotony of the work.

8

— — — ‥

Up the Track

Ed had been an equally willing recruit. But by now, as well as flu, he had a hacking cough, his back was covered in mosquito bites and he could hardly breathe in the forty degree heat. Like the men a hundred and twenty-five years before, he worried that there were no doctors for the next 2,000 miles. The one aspect Todd had overlooked in his preparations was the need for a doctor. After six months, Dr Renner was appointed in Adelaide at a salary of £500 per year. Unfortunately, it was too late for Beckwith, leader of Section C. Suffering badly from sunstroke, he had received little sympathy from his men. When he was finally relieved, he could barely journey back on his horse, and died the day he reached home. Dr Renner found a couple of cases of scurvy and an Afghan camel driver who had been ill for a week with a snakebite. But if a man broke his collarbone, it was too impractical to wait for the doctor, and the overseer would set it himself.

Arriving in the small frontier town of Quorn in the dark, Ed was barely conscious. Another sleepless night coughing away on the ground might put an end to our trip. The one motel was shut for renovations and people were closing the curtains of their clapboard houses. Only the Pichi Richi takeaway was open, selling fish and chips, smothered in batter, 'or would you prefer spam fritters or sausage in batter, or a battered Mars Bar? You just coat them in my special recipe and drop them in the pan. The children love them.' The woman chattered away, amazed that anyone would visit Quorn in the heat of the summer. 'You'll ruin your skin,' she said.

Having politely bought a couple of cans of Coca-Cola, and a home decorating magazine, I asked whether there were any rooms in town. The woman knew someone who rented out houses to labourers and the occasional tourist in season; we might be allowed one of those. An hour later, my hair now saturated in cooking-oil fumes, a boy appeared on a motorbike and told us to follow him. He led us to a house in the outskirts and unlocked the door, asking for his dollars straight away. 'Mum said you can have these for breakfast, seeing it's meant to be B&B,' he added, handing over a basket of eggs, milk, bread and butter.

The house hadn't been used since the last harvest. Dead flies were stuck to the cooker and an old mug of hot chocolate was now foaming with green froth on the side. Angry flies buzzed in the bathroom and the loo was dirty from the last inhabitants. I deposited Ed on the sofa. The house was stifling with the shutters closed. There were three bedrooms and I didn't dare inspect the sheets. Deciding to make scrambled eggs proved a mistake. The first egg was fertilised. Feeling queasy, I managed to mix the next five with the milk. Then, unable to find a frying-pan that wasn't inch-thick with grease, I shoved the mixture into a bowl and threw it into an ancient micro-wave. The eggs puffed up and I added two slices of stale bread, but the butter was rancid. I binned the concoction. Next door, music was blaring from a radio. I put Ed to bed with three sleeping pills, the only medicine we had.

The rest of the night was a blur. The house grew hotter and hotter as Ed spluttered into his pillow, shuddering convulsively and scratching at his swollen bites. I tried reading, but the lights were soon black with mosquitoes. Around two in the morning I ran a bath. The water was brown, and I was sick. Stumbling outside, I was growled at by two dogs circling the house. I picked up the telephone. It was dead. As soon as it was light, I bundled Ed back in the car.

Quorn in the daylight looked surprisingly quaint, with its one gas pump, its coffee shop, stock-agent's, butcher's and stone town hall. Seersucker dresses were sold in the hardware store, alongside dungarees, rolls of gingham cotton and paper doilies. Plastic tea sets were squeezed between a washing machine and a lawnmower. A man came in to buy two dolls' outfits for his daughter's Christmas present, and a rechargeable hairdryer for his wife. The grocery store was surprised when I put the yoghurts back because they were a month past their sell-by date. 'No one round here eats yoghurt,' the woman laughed. The café sold feta-cheese pies, but the town only needed a few bullock wagons and Gary Cooper could have been waiting for High Noon by the disused railway station. After a few hours sitting on the café veranda listening to old ladies with large brooches gossip about who was going to be next year's May Queen, and being fussed over by a long-legged young waitress, Ed revived slightly and we turned on to the highway for Hawker.

Hawker was eerily quiet. 'They've forgotten us up here,' said the shop-keeper, fixing us ham sandwiches for tea. 'South Australians are the poor relatives in this country and we're the black sheep of the family. We've only got one TV station and they won't even pay for a decent secondary school north of Port Augusta.' She produced some tattered photographs of women in velvet dresses and men in white breeches. 'My great-grandparents were some of the first settlers up here, wheat farmers in the 1880s, they blazed the trail. But there's too much drought. The last harvest was in 1947. You'd better stock up on some tinnies, it's going to get worse.'

Subdued, we headed up the highway, inhaling the dust, until Parachilna loomed out of the dwindling scrub. This pit-stop had a couple of old shacks, two caravans, a water tanker, and a sign saying 'The Prairie Hotel: Rooms to Let'. From the outside the hotel looked like a corrugated sauna, but inside it had four large freezers and a menu that included lime-wrapped trout with fennel compôte. In the bar it was almost chilly, and we took our stools beside a schoolteacher, two drovers and a trucker. A Japanese cyclist was lying in the corner guarding his bicycle wheels, a wet bar towel over his head. The barman explained that the man was trying to make it from coast to coast living on rice cakes and goannas in preparation for a he-man contest. But he'd already succumbed to a toasted sandwich. 'It's something to do with the sudden cold; this air conditioning does funny things to people,' the barman said.

A French film crew had stayed at the hotel while shooting a science-fiction movie and had installed the highly prized air conditioning. A man arrived from the nearby railway line to tell us the temperature outside was forty two degrees. His job was to check the thermometer every day and, if it got above forty degrees, to phone Port Augusta because it meant the sleepers risked buckling and the daily goods train could be thrown. 'Wouldn't it be more dangerous if the passenger train got derailed? ' I asked. 'We haven't had passengers coming up this route for years,' he said, looking at me as though I was demented.

The Prairie Hotel was founded in 1870. Jane, the current owner, whose husband ran a property up the road, bought it as a shack when she'd got bored of cooking corned beef for the men every day and realised her children would soon have to be sent south to board at the nearest secondary school. She enjoyed concocting exotic recipes in the kitchen. The enterprise had been a resounding success, even though it meant she had to drive a hundred miles a day to and from work. The locals ate the home-made pork scratchings, but never touched the duck in redcurrant sauce. Truckers were served the best steaks before Alice, and tourists came from Adelaide after hearing about the clean white sheets. Occasionally, Jane organised a band and a barbecue and the nurses would arrive from Port Augusta to spend the weekend partying away from their patients.

We spent the next couple of days at the bar while the Japanese man bathed his feet and the schoolteacher marked some homework. The conversation kept returning to the issue of lowering the age of consent for homosexuals, which the drovers thought was disgusting and the school-teacher was defending. 'Only another five months, and it's back to Adelaide and civilisation,' the teacher muttered. When we explained our plans, he looked concerned. 'Don't get taken in by all that romantic explorer talk. Those men ruined this country. They slept with Aboriginal women, gave them grog and stole their land. Your history books tell you only about nice white history, Alfred burning the cakes and all that shit. You have no idea what white men did to this country and are still doing to it.' Unnerved, we promised to pick up our litter as we went.

Jane pointed us along a dirt track in the direction of Old Beltana. This was the first surviving repeater station on our list. Having taken advice from the American telegraphists, Todd realised that the telegraph signal wouldn't be nearly powerful enough to continue unaided from Darwin to Port Augusta. So at intervals along the line, he planned manned outposts, where men would receive the Morse-code signals and tap them out to the next station down the line. Each station would be made up of a receiving office and two four-roomed houses with glazed windows, white fences and a plot of land to grow vegetables. The job would be bleak, with only the messages to keep the stationmasters amused.

The stone buildings in Old Beltana were still standing, and there was a faded sign claiming them as part of South Australia's heritage. But they were not an obvious tourist site. Signs on surrounding shacks warned intruders that large dogs would attack them, with one reading 'Dogs Beware Thieves'. Neither the promised canines nor their owners were visible, just a solitary row of washing. Once this had been a hippie commune. Today, peace and love had given way to reclusive mistrust and the atmosphere was eerie. New Beltana, ten miles away, is where the local rednecks go to down a pint and bet on the horses once a year at a makeshift racecourse.

The top end of the South is peppered with abandoned settlements. Another fifteen miles east, along a potholed lane, is Blinman, the highest town in the state. Its front street is tarmacked and next to one ruin there are worn remnants of a tennis court. Only the graveyard hints at the fact that 2,000 men once flocked here to mine copper. The mine closed in 1918. Today, the names have been sweated off the gravestones and large black crows circle overhead. A handful of residents still testify to the dream. One runs a deli for tourists in the winter months, another works as a car mechanic for local properties.

Turning back along the crest of the Flinders Ranges, we couldn't make out a single hill between us and Darwin. For many day-trippers, the rolling Flinders with their turquoise parrots epitomise the outback. They go to Wilpena Pound, a bowl in the hills named 'cupped hand' by the Aborigines, surrounded by sheer cliffs and high peaks. But we were heading for the flat ochre tablecloth of the interior.

Meandering along dirt tracks, we passed dried-out waterfalls and abandoned cement mixers. By now we were used to the solitude. Then, rounding a corner, Ed almost ran into a naked figure, and squealed to a halt in front of a shiny new Range Rover.

Nervously we looked back. The woman was still there, by now draped in a towel and being hugged by a photographer. Louise was doing a fashion shoot for a new men's magazine, which she hoped would launch her modelling career. Her props included not just the car, but an Akubra hat and Jake, who had a long black ponytail and was also naked, except for a pair of Wellington boots. 'Other magazines never do anything original,' the photographer explained. We agreed that the couple looked amazing in the dappled sunlight, their brown bodies framed by the white bark of the gums. Jake had donned some leather chaps, Louise was posing as a naïve prairie girl. Ed and I declined to be extras, but we promised to buy a copy when we finished our trip.

Back at Parachilna, a back-up crew had arrived on its way to help a runner break the world record for crossing the 235-mile-wide Simpson Desert in the height of summer. He had to beat 3 days, 18 hours, 52 minutes and 18 seconds. Having tried a few dawn runs along the tarmac and felt the sweat dribbling down my face, I knew he must be an insane self-publicist. The crew admitted their man was prone to leg cramps, and at the last attempt had lost ten kilograms from dehydration. The wide open spaces seemed to attract nutters and schemers. The next grand project for the area was a vast 'Opera in the Outback'. The locals planned to fly 10,000 people from around the world to listen to sopranos singing in the desert night. There were only fifty tourist beds in this border country, so a thousand tents would have to be pitched and five hundred portaloos trucked in.

We set off past Leigh Creek, an old coal-mining town, now a dilapidated prefabricated warren. It was covered in 'For Sale' signs. The pistol range was closed and the golf club abandoned. This town would soon be another Beltana, but the nutters would survive. Talc Alf, a man who made carvings of animals, lived in a caravan half an hour up a red track. He was sitting in his black-and-white checked car, listening to a tape of the Goons when we arrived. What was he doing here? 'The only true wilderness is between a greenie's ears,' he replied.

Hergott Springs, now renamed Marree, was the last stop before the tarmac faded out. A large hotel dominated this old intersection of rail, road and telegraph lines; the saloon was empty, as was the police station. Once Marree was home to sixty Afghani cameleers and their families who had their own mosque and 1,500 camels. Now the only camels come on key-rings and are lime green. The Ghan Store made us chips from the freezer. A local was talking about swiping some old sleepers to make a kitchen table before they had a preservation order stamped on them. Emerging into the dazzling daylight, we realised we'd been eating lunch next to an original repeater station.

At Marree, we turned on to the Oodnadatta track, a wide dirt road griddled with corrugations from the cattle-trucks which used to hurtle down south in winter. After five hundred miles, we were finally leaving the tarmac behind. We resigned ourselves to the impossibility of rescue should our vehicle break down, and set off into the vast, white wilderness scorched of all life. For the next 130 miles, the only settlements would be ruins.

At first we bumped along slowly, but our coccyxes were soon screaming, so we tried a more direct approach, driving as fast as we could, which was equally bone-shaking. Alternating between a crawl and a dash didn't work either. So we adopted an even pace and tried to surf over the bumps, a technique we later discovered was known as 'floating'. Our destination was the Peake, the furthest point of human habitation in the 1860s. From there we would set out for Alice Springs.

Leaving Adelaide in October 1870 with the last of the teams, Todd spent five weeks inspecting Bagot's work up to Hergott Springs. The contractor's men had already planted poles one hundred miles north of Port Augusta, and had cut enough trees to reach as far again. It was now mid-summer and the earth was almost impenetrable, but Bagot estimated that he would be finished in nine months. Todd was jubilant. Returning to Adelaide two days before Christmas, he thought the line might be finished on time.

Alice was relieved to have him home after three months up the track. But she knew she came second to his work. There was little she could do to help except tour the tea parties, telling smart Adelaide how well things were going with the project. In 1869, Alice and Charles's fifth child had been born into a house in chaos. Todd was already preoccupied by the telegraph, and, according to Pat, Alice had been obliged to rest in bed for much of the pregnancy and then suffered what would now be diagnosed as post-natal depression. Mrs Hall called it 'nerves'. She insisted on coming to stay, and Alice, still feeling guilty about those first few days in Australia, felt too weak to say no. When my great-grandmother was born, Charles was so obsessed by preparations for the line, he barely acknowledged her arrival. The observatory was in disarray. Visiting afternoons dwindled, and few came to call. No one thought of christening my great-grandmother, so the children nicknamed her Nina. Alice spent more and more time reading novels. At thirty-three, she had begun to accept her role as an increasingly stout matriarch, but still slipped out into the garden to read novels under the pepper tree.

There had been several more additions to the family over the years. Griffith Todd, Charles's seafaring elder brother, had 'gone native' out in Bengal. His first wife had died, and he'd married twice more, before succumbing to swamp fever aged thirty-six in Calcutta. His daughter,

Fanny, had returned to Adelaide with Alice after her visit to Cambridge. Blessed with a 'sunny disposition', according to Todd, she looked after the younger children, helped run the household and provided company for Alice. Her two brothers, George Griffith and Charles Robert, soon joined their sister, to be looked after by their 'rich' uncle and aunt.

To complicate matters further, four years after Alice's father had died in 1865, her mother had insisted on coming to live with her adored youngest. Mrs Bell was horrified by the ill-disciplined brood she discovered. Having first become firm friends with Mrs Hall, she spent her time chastising Charles for his messiness and her daughter for her lack of spirit. But she had at least brought with her much-needed money. When her husband died, his capital had gone to the eldest surviving son, Edward, but Grandmother Bell had been given the interest on it, a sum of about £400 a year.

The building of the telegraph was to prove the severest test of Alice and Charles's relationship. But in that Christmas week of 1870 these troubles were still in the future. When Charles told Alice about his initial success, she was thrilled to be able to report to a party at Lady Bacon's house that the work was running smoothly. They would both come to regret their optimism.

9

— — —— ·

The Singing Line

Since Stuart's trip seven years earlier, no one had ventured further than the Peake. Todd's advance exploration team had to find a route to Darwin before the central sections descended on the new depot with their horses and stores. The first stumbling block was the MacDonnell Ranges. Stuart's men had scrabbled their way over the hills; Todd was going to need a proper pass for his poles. The other logistical problem was water. Cutting trees and digging holes was thirsty work in the outback, and the horses needed water as well.

South Australia didn't have the legendary drovers of New South Wales and Queensland, who pioneered the search for more grazing land. Few men in the South became 'kings in grass castles', sweeping cattle before them in vast arcs. Two-thirds of this new colony still lived in Adelaide, and the best steaks came from small coastal properties.

Todd asked those who had been brave enough to push north in search of pasture whether they had any good bushmen who could make up an advance party. John Ross, manager at Umberatana and Beltana, was the name they repeated. He was a Highland Scot who had been a drover in South Australia since he left Scotland, aged twenty, in 1838. At least a foot taller than Todd, Ross had arrived in the new land looking like a Florentine prince, with curly black hair and an imposing forehead. Now his head was bald and his beard was long. He hadn't yet travelled as far as the border with the Northern Territory, but he was the only man who knew anything useful about the country beyond the salt lakes. Todd made him leader of the expedition at £450 a year, trusting him with the project's future. Ross had less than six weeks to find a tree-lined route through the centre, with plenty of water, but not liable to flooding. He must have suspected that Todd's schedule was insane.

As the explorer assembled his provisions in the York Hotel in Adelaide, a young man called Alfred Giles sought him out and begged to be allowed to join the trip. 'Are you sound in mind and limb? Can you live on bandicoot and goanna?' Ross asked. The other members of the party weren't so robust. There was William Hearne, a stock-rider who was lame in one leg, and Tom Crispe, a bushman who had never recovered from getting lost for nine days on an ill-fated exploration to the north. The surveyor was William Harvey, who had reached the Northern Territory with George Goyder's exhibition in 1869, but 'could barely sit on a horse', according to Giles. This motley crew was offered £1 a week each.

Four days after Ross's appointment, the new team caught the train to Burra. They stocked up at Oweandra, near Quorn, buying twenty-two horses, and were so well entertained that Ross overturned the buggy and was unconscious for two hours. At Beltana they assembled their gear, and at Mount Margaret, the nearest station to the Peake, they plaited whips and picked up three hundred pounds of smoked beef and a hundred pounds of 'jerked' beef, made by cutting the meat into thin strips and hanging it on a wire fence for a day until it shrivelled up.

As we set off along their route, I read out passages from Giles's diary, *Exploring in the Seventies*, an account of his pathfinding trip with Ross. In the introduction he wrote: 'In 1870 the furthest outpost of civilisation north of Adelaide was the Peake Station. Between that point and Port Darwin there was not a sign or mark of horn or hoof except the faint one left by Stuart in 1862, and with the exception of savage tribes and a sparsely populated coastal fringe, an empty continent.' Giles was thrilled. 'The fact that we were the first white people to pass through this country greatly added to the pleasure we experienced.'

The men may have looked weathered, but they were romantics. 'What possibilities these unknown lands might possess of undiscovered wealth, of great rivers, of bold mountain peaks, canyons and waterfalls, and of new minerals, animals, plants and blossoming forests,' Giles wrote. 'Not least, there will be wild tribes of savages. We might have battles, and fight our way to the Indian Ocean.' There had been no time to gather proper equipment; the waterbags were made of poor canvas, their horses' girths were not leather and they had no tents. They did have several tomahawks to present to 'chiefs of wild tribes', and Ross took a red flannel nightshirt to wear to bed.

Within weeks, the men's legs were being shredded by the porcupine grass and they were surviving on wild turkey and duck, which they wrapped in mud and buried under their campfires to cook. After a couple of hours they could peel off the clay and the feathers would be stuck to the mask. Pat had showed me how to do it at her property, using one of her chickens as a substitute. 'It's even better with hedgehogs,' she told me as I watched the dead hen being smothered with gunge. 'The gypsies in Eastern Europe used this trick all the time.' When the birds petered out, Ross encouraged the men to eat boiled salt-bush and the occasional owl's egg, as their provisions were already pathetically low. 'However, some of the flowers were of the most beautiful description. Hearne and I decorated ourselves and horses every day with bouquets,' Giles wrote.

Time was wasted as they trekked to higher ground in rainstorms or looked for waterholes after days without drink. After one thirty-six-hour stint without water they saw smoke ahead. 'Where there is smoke, there is fire, and where there is fire there are blacks, and where there are blacks there is water,' Giles wrote. They soon discovered an Aborigine carrying some cooked goannas. In the distance they saw magpie and black cockatoos, but for another day and a half they could find no water. Turning back, they pleaded with the Aborigines using sign language, and were led to a waterhole after seventy-two hours without a drink. Of their first encounter with Aborigines, Giles wrote: 'Shortly afterwards a fine-looking fellow marched boldly into our camp with his hair in ringlets and tied at the back of his head. He was perfectly naked . . . Before night their lubras [women] came into the camp. To one old woman, whose ugliness it is impossible to describe, I handed a looking glass.'

Todd had given strict instructions to the forward party that they were to avoid a fight with 'the natives' if at all possible. The Aborigines were an unknown quantity in his plans. He knew he couldn't afford to tie up any manpower in a war along the line. Ross's team humoured the Aborigines at first but when they started stealing food by hiding it under their armpits and hair chignons, and smoked the Europeans out of their camps with bushfires, Ross fired a few shots in the air.

The instruction manual to the overseers was explicit. Under treatment of natives, Todd wrote: 'Should any natives be met with, they must be treated kindly but firmly. No one is to be allowed to visit the natives' camp without special permission and in all cases previous intimation is to be given to the natives.' The next command is written in bold. 'No communication with native women'. Todd adds that this could only stir up jealousy among the Aborigines. He also insists that the 'natives'' property and burial sites should not be touched. 'They must be warned by cooeying of the approach of a white man, as their first impulse of terror at the unaccustomed sight often leads them to throw their spears at him.' He ends the section insisting: 'It is most strictly forbidden to fire upon the natives except in the last extremity.' Many men took notebooks and scrawled down Aboriginal names for words such as track, food, water and fire in case they got lost.

We wondered if we would find any Aborigines who had heard stories about the construction of what they later began to call, 'The Singing String' or 'The Singing Line' – the piece of wire that cut through their land like a cheese slicer. In one move, this wire brought to an end thousands of years of unchanging nomadic existence.

By mid-September the forward search party had found a dry creek sixty miles east of what is now Alice Springs. The MacDonnell Ranges loomed ahead. There were plenty of pigeons to eat but they had no water left. They turned for home defeated. Giles wrote in his diary: 'See that cliff up there? If it were solid gold I would not exchange it for this quart of water.' Stumbling across Chambers Pillar on their return, they carved their names in the rockface of this soaring red tower which had first been discovered by Stuart.

Continuing south, they reached the Finke River, but found it salty. They followed it twenty miles to the junction with the Hugh River, and there finally discovered a small pool of good water. This soon became the central section's well, nicknamed Alice Well after Alice Todd, who had sent the men more books, a consignment of new shirts from the ladies of Adelaide and some Dundee cakes. The telegraph absorbed all conversation for those stuck back in the city, and everyone was trying to do their bit. But at this stage, Ross and his men had no such luxuries. They were living off damper, a bush staple made of flour and water, and had run out of tea. When they limped back to the Peake depot to tell their stories, they estimated that they had covered 1,000 miles in nine weeks. 'The total absence of any corpulence on our bodies convinced them of the truth of our assertions,' Giles wrote.

He knew the project was in trouble. 'From our experience, so far as we had been and seen of the interior, it meant facing tremendous and unforeseen difficulties. It was impossible to guess when and where the two ends of the wire would be joined.' Since the end of September, the five central section parties, including fifteen officers and eighty-seven men, had been making their way up the line. The Peake, once a solitary stockman's outpost, was soon a canvas city. Bacon and Blood arrived with their stores, the one hundred Afghans followed on their camels, and unloaded the wire. They were led by Hadjee Meer Ban, dressed in pure white robes, which he changed twice a day. On 10 November a buggy arrived and a dusty man stepped from it. It was Todd. Hearing that Bacon was 'quite lost when he comes to pen and ink', Todd made an inventory of the storerooms before holding a church service for the men. The next day he settled down to discussions with Ross.

The Scotsman was eager to paper over any difficulties. They agreed on a route to the Peake, but Ross couldn't bring himself to admit that he still had no idea how to cross the MacDonnells. Todd had to return to debrief parliament, so handed control to Overseer A. T. Woods. Before heading back to Adelaide for Christmas, Todd praised Ross's good work over a rare glass of beer, upped all the explorers' pay, and asked them to redouble their efforts. As Ross set out in the mid-summer sun of December 1870, he realised this trip would be even more testing, and had little hope of success. Todd's enthusiasm was almost too much to bear.

10

— — — — — — — — —

Red Dust

Ed and I were not so naïve about our next step. Our swags were gritty with dust, and we knew that there was only one opportunity for a cold beer in the next two hundred miles. Driving westwards over undulating gibber downs – flat plains covered with wind-polished stones – we soon approached the great salt lakes. Eyre North and Eyre South were a shimmering expanse of white-blue glass. Stiff peaks, formed by the winds, seemed to lap against their shoreline. This 6,000-mile-square lake-bed was a mirage of our final destination, the Indian Ocean. It has filled to the top only three times in the past 150 years, when pelicans and seagulls flocked to its shores.

Taking off our shoes, we dabbled at the edge, threw stones and tried tasting the salt crust. We soon burnt our feet on the scalding white rim. Cooling them in a reed-fringed waterhole that bubbled with sulphurous belches, we watched a lone duck circling overhead. To Ed's amazement, I threw myself into this sandy cauldron and disappeared, only to float back to the surface on a jet of gas and sand. 'This,' I shouted to Ed, 'is a blubber pool; you can hurl yourself into the centre and the pressure of the gas lifts you up.' I'd read about it in Giles's diary. Ed and I spent the afternoon in our jacuzzi.

Covered in sticky sand, we returned to our Toyota. A turquoise butterfly landed on our bonnet. Two miles further on, we nearly ran over a crimson desert rose that had just bloomed. We had arrived at Curdimurka, named by the Aborigines after the kadimakara, mythical beasts thought to live under Lake Eyre, preying on anyone who dares to walk across it. The next day, by a rusting water-tank, we turned off the track and headed out over the dunes.

When Anthony Trollope took a day-trip to the edge of the outback in the 1870s, he wrote that it was 'brown, treeless and absolutely uninteresting'. Yet this great space is, for a visitor from the northern hemisphere, the strangest aspect of Australia. The cities of Adelaide, Sydney and Melbourne provide reassuring echoes of the old world, with their cathedrals, parks and plazas. Their hinterlands duplicate Britain's countryside, smothering the native flora with daisies and taming the landscape with paddocks and hedgerows.

The outback is too hot to grow any disguises. Europeans called it the new world, but it is an old continent. When you fly overhead, the endless flat, burnt plains seem unchanging. On the ground, every mile is different, either a subtle change in the stones or the vegetation, or a surprising colour or shape. One minute the bleached scrub stretches ahead, the next a white crane flies over a desert oak or a flock of bronze-winged pigeons settle on a hill. Only the size is unchanging, big skies a constant canvas against which details are delicately painted. The complete composition has a minimalism that appeals to modern tastes. It repays attention, rather than surrendering to a casual glance.

We could have driven round this wilderness for days, but we were running out of fuel. If we broke down, it would be weeks before anyone noticed we were missing, and days before they spotted the red flag on our aerial from a plane. 'Tourist dies in outback trek after car breakdown' was the last headline Ed and I saw in the *Guardian* before starting our trip. An Austrian medical student had died after her Toyota had become stuck in a sandy track near Lake Eyre. Ed told me sternly that she had overruled her boyfriend, who had tried to persuade her to stay with the car, and had set off alone. She was found dead two days later by the publican of William Creek. The constable from the Port Augusta police had said that many foreigners 'chose to underestimate the dangers of the outback, in some cases from a perverse sense of achievement, and some have perished'.

We drove gingerly from the lake shore until we came to an empty dam, built for the steam locomotives by the Edwardians and now used only by the screeching corella cockatoos. Heading for the Peake, South Australia's furthest outpost in 1870, we passed a pole. This one had gone 'feral', as the Australians say about anything that turns wild. One of Bagot's metal poles, it had lost its insulator and its wires were flapping like tentacles in the breeze. We were still impressed, and walked round it, examining its scratches like some mystical totem pole, and photographing it from every angle.

The Peake was above a rocky hollow, overlooking the valley and surrounded by cork trees. While Beltana had felt oppressive, this telegraph station was peaceful, the air cool and we felt a few drops of rain. Nine white stone buildings were still identifiable. We found a tortoiseshell hairslide and a horseshoe, and identified the storeroom and the manager's house. Sitting on the piles of rubble, we finished Giles's diary. Two hundred men camped here on their way up the line; today their traces have been obliterated. In the remaining light, we bumped back on to the track and returned to the only settlement in this wilderness, at William Creek on the saltbush plain.

'Hurry up or you'll miss the bar snacks,' an old man shouted as we walked into the pub. Oysters in hot chilli sauce were lined up on a plate on the bar. He took us to the caravan to inspect our beds and have a wash. 'Supper's at seven, don't be late or my daughter'll get twitchy. No noise after nine p.m. or you'll wake the grandson. The generator goes off at ten p.m., no cooked breakfast after nine p.m. The drinks will be waiting.'

The pub was built in 1887 for the camel-drivers working the overland route from Adelaide to Alice. Grey T-shirts from around the world hung from the ceiling, and a huge bra filled with dollar coins swayed from the rafters. 'Bet you'd like her address, but I can't give it to you.' The grandfather swung the bra and winked at Ed. Bottles of dried-up liqueurs lined the bar with a jar of pickled eggs and a snake in formaldehyde. It was forty-two degrees. Water was the only drink they didn't have.

The ceiling fan turned half-heartedly. In one corner the local grader, paid to flatten the road's corrugations, was engrossed in a fuzzy version of *Pretty Woman* on the video. He was being ignored by the station-hand and his wife on their Saturday night out. Women in the outback often seem very white, very frilly and very put-upon, like the leading lady in a Clint Eastwood movie. This one was no exception. She was dressed in a pink smock, with a bow in her blonde ponytail, and she looked exhausted. The straps dug into red raw shoulders and two toddlers hung off her freckled arms. She could have been anywhere between eighteen and forty.

The pub was a family affair. We had passed a plastic swimming pool and trampoline on the way in, and the publican's children kept ponies in the backyard. 'Steak sandwich or whiting?' the grandfather demanded. We asked the advice of the only other tourist, a pathologist's secretary from Sydney. 'Oh, I wouldn't have the fish. Had some the other day and nearly killed myself,' she said, bringing over her bar stool. 'It was a can of tuna and I'd left it in the sun in the back of my car. It smelt a touch sweet but I was so hungry I ate it with my fingers. Started swelling up all over.' She pointed to her throat and mimed a throttling action. 'Couldn't breathe. I was ill for two days, so blue I looked like a mortuary photo. My fault, my mother said don't eat fish unless you can smell the sea.' We ordered the steak with tinned salad and took it outside to watch the sun set over a small plane, parked casually next to the lone petrol pump.

The publican joined us. 'We answered an ad, threw everything in the boot and just drove up here,' he explained. 'That was two years ago. One hundred tourists a day can pass through here in the winter. We're a speciality destination. When we took over the place we tried to smarten it up a bit. But the tour groups like it authentic, so we serve them their beef pies and white sliced bread. I prefer moussaka myself.' We realised that the unforeseen advantage of driving up in this heat was that we had missed the tour groups.

Wasn't the work shattering? 'You can make a killing, but you kill yourself in the end, telling the same old stories sixteen hours a day,' the manager replied. 'And there's nowhere to escape. Finding staff is impossible. Kids come from the city, their electric razors don't work and they go home again.' His wife wished they'd lived here earlier in the century. 'There was a chain of small towns then and there was a real community. Everyone met for christenings and coming-out parties. Now we're a seven-man town excluding the children.'

Her husband agreed. 'Ten years ago the station-hands wandered in here covered in mud and blood from chasing the cattle. They'd been in the saddle a fortnight, and could down twenty beers. Now it's a couple from Kansas City sharing Diet Coke.' In one house there was a German banker who played the stockmarket all day on his computers, next door lived two historians, the pub employed a tutor and then there was the grader. The grader's task was Sisyphean. Every time he finished smoothing out the local roads he had to start at the beginning. No wonder there were six empty bottles of beer in front of him. He had left the surrounding properties only once in fifteen years, and that had been a mistake. He'd been enticed to Coober Pedy to have a burger. 'It's primitive here, but it's worse there,' he said. 'They live in caves and hit each other with rocks.'

According to our schedule, we didn't have time to check out Coober Pedy, the toughest mining town in Australia, home to anyone insane enough to try their luck with a shovel in search of opals. The name means white man's hole in the ground, which is hardly a tourist slogan.

Then Ed started coughing again. He drank two glasses of hot whisky but his wheezing kept everyone up all night. The next morning he was sweating profusely and had come up in a rash. The publican's wife took his temperature. It was 103 degrees. 'You just need a doctor and some rest,' she said. The flying doctor service was shortstaffed over the December holidays and Ed wasn't yet an emergency. Alice Springs would take three days to reach going flat out. Port Augusta was days behind. I was so annoyed that I went to cool off in the plastic plunge pool. Sitting in the tepid water, I worked out we would have to drive to Coober Pedy after all, a six-hour detour along a sandy, corrugated track. With a population of 10,000, this inferno must have a doctor. I've never liked opals – their speckled pinks, oranges and yellows are too glitsy and they're bad luck. It was only two weeks to Christmas. I tried tempting Ed with the Christmas stocking I'd brought from England. He said he'd open it later. So we got in the car and drove west.

Meandering slowly through the dusty outback is a soothing experience. Driving at full tilt, sand spewing in all directions, careering from pothole to ditch, is shattering. Ed would momentarily revive enough to start arguing that I should be wearing a seat belt. 'If we meet a car, it's plastic surgery time,' he said. We hadn't met one yet. We didn't stop to appreciate the dog fence, the largest stretch of wire in the world, which goes from east to west across Australia, penning the dingoes in the north so the sheep can graze in the south. Ed fell into a fitful sleep.

On our left was prohibited territory, atom bomb-testing country. As we finally approached Coober Pedy we couldn't see any sign of life, but we'd been told that everyone lived underground because it was cooler. A large sign, circled in red, showed a man walking into a deep hole. We passed the digging machines and came across mound after mound of spewed-up red earth. They looked like molehills on an English lawn, ruining the view for miles around, so you wanted to stamp on them. Carcasses of cars were strewn everywhere, machinery rotted at the side of the road and lorries marked 'Explosives' careered round the bends. 'Seat belt', Ed croaked.

Dugout caves were slowly being abandoned in favour of houses with fans, and some even had foundations. For the first time in a week we were back on tarmac. The whole town was linked like a dot-to-dot picture with wooden telephone poles. The main high street was lined with chalk-board signs announcing the opal buyers. Fifty-one different nationalities were meant to coexist in Coober Pedy. We checked into the only three star hotel

I could find. It promised old-fashioned underground caves, video recorders and room service.

The receptionist pointed us to a hospital on a hill. At first it looked empty. Then we saw an Aboriginal woman walking out, a drip machine attached to one arm and a bottle of whisky in the other, to chat to a friend. Another Aborigine, her arm in plaster, came to join her on the steps and an old man came out in a wheelchair, stood up and walked off down the road. Three Aboriginal children were being handed sweets by a nurse as we went inside. The nurses had created a small oasis filled with ferns in the middle, 'To keep us all sane'. It was the only garden in Coober Pedy. We waited outside the doctor's room, watching the woman on a drip slowly drain her bottle of whisky. The coffee machine was broken. The ward beds were empty; miners can't afford a day off work and the Aborigines refuse to stay.

Ed's name was finally called and he disappeared. Half an hour later he emerged grinning. The doctor had had a cathartic effect. 'I've got bronchitis,' he said smugly, waving his prescription. 'I'm not going to die, but you've got to treat me very carefully. I need pills, cold Coca-Cola and days of ice cream. There's a video shop on the corner.' I had thought that only pensioners living in rainy climates got bronchitis. The doctor had issued little advice, but he had made Ed laugh. 'He was sitting there with two certificates and he couldn't remember which of his patients had just died and which needed certifying as mad. So he tossed a coin.' The doctor had only taken the job because he was a gambler. He said he'd tried every other game. Weekend opal mining was his last bet.

While Ed watched old movies, I went to the gym, expecting it to be empty in the middle of the day. It was packed with men in tight vests pumping serious machines. The town is full of people who have either lost their jobs, their families or their self-restraint and have come to bury themselves in the red earth. In Coober Pedy no one ever asks for your credit card or driving licence. You just rent a digger on credit, buy a PSPP – Precious Stone Prospecting Permit – take four wooden pegs, mark out your ground and start sifting through the earth, making sure you're not trespassing on anyone else's patch. Of course there are tensions – the police station got burnt down one year and there are fights on Saturday nights – but mostly they understand each other's obsession.

The women who followed in their wake have given up hope of riches. They've either got jobs in the tourist shops or congregate at the hairdresser's and spend their evenings at the drive-in movies on the edge of town. None of them wear opals, although you can buy $10 earrings and necklaces at the garages. They import their own luxuries. At the moment there was a craze for fishtanks; last year it had been bread-making machines. Next, the women wanted to set up a water-bottling plant and sell 'Coober Pedy Desert Water' to the masses.

Lucky families were packing up and going away for the holidays. 'Send me a postcard from Tasmania, Perth, Darwin . . .' resounded round the public swimming pool. It was like a boarding school, but the only lesson here was that life is unfair. The receptionist's father was a digger; he'd come to try his luck twenty-five years ago. Fifteen years later he'd found five huge opals and the family went back to Holland. But his wife ran away and the money ran out, so he came back with his daughter and started again. He just hadn't struck the right seam. The receptionist asked if we'd like to go 'noodling'. I must have looked surprised at what sounded like an invitation to foreplay. She explained that all you had to do was get down on your hands and knees and fossick in the dirt for gems. It sounded exhausting, so I opted for a gentle stroll, while Ed went back to his three-star cave.

A church was dug out of the sandstone in 1977 and has proved surprisingly popular. Even the ministry appears in thrall to the opals. 'As miners struggle in oppressive heat and isolation, dig, gouge, sweat and blast to bring the opal's beauty out into the open and gain a rich reward, so also God is at work, seeking out gems for his kingdom,' says the sign by the door. Another reads: 'Just as the opal buried deep in the earth fails to display its natural beauty, so too does the man buried in sinfulness.' Either the vicar was a part-time miner or he was a pragmatist. A tattered sign to 'underground bookshop' drew Ed's weakened attention. But here 'underground' didn't mean avant garde, just cookery books – more risotto than risqué.

After a few days of eating bercher muesli in the Swiss café over my *Coober Pedy Times*, working out at the gym, hanging out at the hairdresser's and watching grown men build sandcastles, I began to relax. In the afternoons I drove the Toyota to a drive-in movie in the desert or read the classified ads. 'Sarah is still single, despite angelic features, long golden hair, brilliant blue eyes, porcelain skin with natural rosy cheeks and a gorgeous smile.' She desires to meet 'a down-to-earth gentleman', which I thought should be easy to find in this town. Ed scanned the police reports. Next to 'fight at Serbian club' he smugly pointed out the chilling words: 'Numerous persons cautioned – fined for not wearing seat belts'. I got my hair cut and my toenails painted 'sunset pink'. No one questioned why we were in Coober Pedy and we couldn't summon up the energy to leave.

Ross's team was having different problems. As they set out on their second trip, they were accosted by a native. The native gabbled that he had been killing blacks and asked them to shoot some more for him. The men refused, but were worried that it was a bad omen as they redoubled their efforts to find a way through the MacDonnells. Soon they were at war with the ants. Regiments invaded every nook and cranny and they had to spread hot ashes across the paths to their tents. Next a death adder wrapped itself around one of the party's foot. By mid-December 1870, the group had finally found some better land, with emu a welcome addition to their diet.

The explorer Stuart had become so blind while making his historic crossing that many of his maps were virtually useless. By New Year's Day the men were almost at Barrow Creek, two-thirds of the way to Darwin, and saw orange trees in blossom and swarms of butterflies. But they had circumnavigated the MacDonnells only by making a detour 250 miles north-east. The poles would never be able to follow such a loopy circuit. They returned to search for a gap in the MacDonnells, stumbling on Central Mount Stuart, which marks the geographical centre of Australia. To everyone's surprise, Harvey unrolled a flannel shirt from his swag to reveal a bottle of rum he had been saving for a suitable celebration.

This time they crossed the MacDonnell Ranges at the same point as Stuart's party, Brinkley's Bluff, and confirmed they could never stretch a telegraph wire over those inclines. They began to lose hope. Then Harvey got lost. 'He was of an excitable nature, not an experienced bushman,' Giles wrote. The men fanned out, taking essence of meat to revive him; but his tracks were incomprehensible. They spent two days wandering in circles, searching for broken twigs. When Harvey finally stumbled into camp he found it deserted, and polished off the rest of the rum and stewed wallaby. The others returned, exhausted and tetchy.

New Year's Day passed quietly. 'We hoped Adelaide might enjoy their picnics and pastimes. Ours was a continual one. We could fly our kite higher than they could and our view was unlimited,' Giles wrote optimistically. But they were reduced to scraping dust from the bottom of the empty tea-bag and making sour jam out of the munyaroo plant. Their clothes were in shreds. Giles had no shoes and the copies of Stuart's maps had all disintegrated. Returning to the new depot on the Finke River, they were convinced there was no route through the MacDonnells. The line would never have a chance to sing.

11

·———— ·————

The Source of the Springs

The overseers were livid. The advance team had obviously been gallivanting in the outback, playing at being explorers, talking of sweet wild oranges, naming hills after their children and panning for gold, but had failed to come back with the goods. One of the overseers wrote of Ross: 'His previous explorations have not been of the least service to us. The delay – the whole of which may be fairly attributed to Mr Ross – will tell heavily upon us, and especially the further sections.' The chief overseer, A.T. Woods, was more measured: 'We have but one object, to put up the line . . . Mr Ross does not appear to keep that object in view.'

Ross had done little more than confirm that Stuart's route was impossible. Team A hadn't used Ross's path. Team B needed to know in which direction to head. Men were stockpiled at the new depots on the Finke and at Alice Well, getting sick from the flies and the heat. Ross himself came down with a fever. Aware that Todd would be anxious for progress, Woods called an urgent meeting with three other overseers. Gilbert McMinn and William Mills, two of the section leaders, were deputised to take two parties to find the missing link. Mills took the eastern approach, McMinn veered to the west.

On 17 February 1871, McMinn found a thin sliver through the rock he called Temple Bar, and the following day he discovered Simpson's Gap, a boulevard through the range. He couldn't believe that everyone else had missed it, and was about to rush back with the news, when a dust storm struck. He was stuck for five days sheltering by his new thoroughfare.

Bumping into Mills on his way back, McMinn couldn't contain his excitement. Mills went up to survey the newly discovered route, but at first it looked impossible. There was no water for miles around; none for the teams and certainly not enough for a repeater station. The great central lake was nowhere to be seen. Without water, the MacDonnells were still impenetrable to the telegraph men; like deserted termite mounds, sucked of all moisture.

Then Mills discovered a dry riverbed and, following it down, found pool after pool of clear water. That night he wrote in his diary: 'Numerous waterholes and springs, the principal of which is the Alice Spring which I had the honour of naming after Mrs Todd.' The search was over. The centre had been breached. With an eye to his future prospects, Mills called the pool after the boss's wife. The dry river he named Todd.

The message went down the line by word of mouth. Todd was eventually told that his men had found an easy route for the line with plentiful water. The principal water source, and the site for the centre's repeater station, would be named after his wife. History doesn't relate which impressed him more. Soon the first tents were pitched and Alice grew out of the desert.

Ed and I were also back on track. We returned to William Creek, where the publican's family cheered Ed's health with warm champagne and frozen garlic mushrooms, and we waved to the grader as we passed him forty miles up the road. The red route to the north disappeared over the horizon. I nearly missed our only T-junction and slammed on the brake, causing us to skid across the track and into a rocky ditch. We clambered out and inspected the result. I'd punctured one of the tyres and dented the radiator.

Ed, now fortified by antibiotics and videos, soon changed it while I watched from the shade. The air conditioning had finally died on us, dust swirled through the vents, and it was the last time I was allowed to drive the Toyota.

We had both heard about our next stop, Oodnadatta, from a painter we knew in Devon who'd been captivated by the place in the 1970s. He spent a year in the town, watching the Aborigines, the nuns and the property owners jostling for space in the desert. He'd even seen a tarring and feathering of one woman, insane enough to try to broker a peace between the frequently warring parties.

The town grew when the railroad arrived in the late 1890s, following the line of the telegraph poles. Trains reached Beltana by 1881, Marree five years later and Oodnadatta by 1891. The Great Northern Railway was a budget build, with light rails and minimum ballast. For thirty years this outpost was the end of the line. Camels from the old telegraph project were used to transport the passengers and luggage from here up to Alice. The train trip was meant to take three days, with overnight stops at Quorn and Marree, but frequently had to stop for a week while flooding subsided or the rails expanded during a heatwave. Once the engineer had to resort to shooting wild goats to keep his passengers from going hungry. On other occasions the driver would stop to let the women pick wild flowers. By the time the train arrived in Oodnadatta, no one cared that they were stepping out into an arid wasteland. This tiny outpost seemed a luxurious oasis. They would spend a few days eating vegetables, grown by Chinese coolies using elaborate irrigation systems, and pay extortionate amounts to stock up on beef before the three-week hike to Alice Springs. The women learnt to ride the camels side-saddle and the men talked business.

When the residents of Oodnadatta saw the smoke plume of the locomotive, the town threw a festival. There were tennis competitions, fancy-dress pageants and even swimming races on the rare occasions the river was up. Nuns arrived to minister to the Aborigines and the population swelled to five hundred. But in the 1930s the engineers finally built a rail-route through to Alice and the town lost its significance. In 1980 the line closed for good, diverted westwards to a new route that didn't flood. The water-tanks rusted and most people left. The remnants stayed on to bicker, and dust drifted over the vegetable plots.

The Pink Roadhouse loomed fuchsia-pink out of the dark. The children at William Creek had insisted that it served the best kangaroo burgers in the outback, but it was closed so we went down the road to the only house with lights in the windows. It turned out to be another hotel. The man at the bar offered us a drink. He couldn't sell any alcohol because this was now an Aboriginal hotel and he was their employee. He'd been in this one-street town for only two months and he would stay only until he had shown his employers how to run the hotel for themselves. We were the first guests that week. Our toasted sandwiches came with the obligatory one slice of orange cheese, the plastic wrapping still sticking to the bread like frazzled bacon. The fizzy cherryade was flat. We bought a packet of toffees and retired to our swags in the dilapidated camping ground, where I explained to Ed my plans to patent the 'outback diet' – we had already lost a stone between us.

I couldn't eat any eggs after our bad experience in Quorn, so the next morning the Pink Roadhouse served me tinned pineapple chunks in custard, while Ed ate fried eggs and steak. The roadhouse boasted a pink tractor outside, pink telephones and pink chairs. The two waitresses were wearing Father Christmas hats with flashing lights on the pom-poms, and cut-off shorts. We watched the locals coming to pick up their post and swap fresh manure for newspapers and wondered why they needed sheepskin seat covers in their cars. They must have been sweltering. The co-owner, Lynnie, wore plaits, a black cotton dress printed with lilies, and full make-up. She had arrived in Oodnadatta from Darwin in the 1960s on a hippie trail to India and fallen in love with the sand.

Lynnie explained that Oodnadatta had once been called Angle, because a tilting pole was used just outside the town so the telegraph line could change direction. 'They changed it to Oodnadatta because they thought it was embarrassing naming a town after a pole, but Angle sounds rather like

Angel; it's a lovely name,' she said. 'There's an angle pole standing just outside the town. But we think it's a fake. The original one was probably made into firewood years ago.' We asked about the rival hotel where we'd eaten dinner. 'Do-gooder governments have been trying to hand it over to the Aborigines for twelve years but the blacks don't have the skills or the patience to run it. The two tribes here fight all the time. Every six months a new white manager arrives all bushy tailed to befriend them, and ends up over here moaning.'

That night there were fireworks and a concert to mark the end of the primary-school term. Children dressed as angels and shepherds flitted around eating hot-dogs and smearing their costumes with tomato ketchup. Adults stood nattering beside a pig roasting on a barbecue. Whites and Aborigines didn't mix. Only the Father Christmas was black. 'Because he's fat, lazy and available', one white woman told us to our amazement. Lynnie handed us a bunch of leaflets just before we left, giving details of the tracks radiating from the roadhouse. 'We wanted to encourage hippies like us to come to the outback, but backpackers kept getting lost in the desert, so we printed these idiots' guides,' she said.

The angle pole could have been any two wooden poles bound together, but we paid it a pilgrimage. There was something appealing in the idea of messages suddenly having to make a sharp right turn. We imagined inattentive missives veering off the edge, as they negotiated the bend. Were there still birthday messages and government contracts lost in the red earth? Did anyone notice if a few dots and dashes vanished?

Not wanting to risk Ed's lungs with too many nights in the open, we headed on to Dalhousie Springs and Mount Dare homestead, where the owner, Rhonda, ran a B&B. Our route took us across some of the bleakest country we'd yet encountered, a stony desert with no vegetation, no animals or birds. The ground was covered in sharp rocks, slippery black in the heat.

It was a relief to exchange this unremitting country for the softer, greener creeks approaching Mount Dare. 'To be truthful, I'll enjoy the company,' Rhonda explained as we arrived at her two-house town at the top of a hill. In one building she ran a shop for the Aborigines and tourists in season; in the other she lived with her husband and dogs. The views across the Simpson Desert had enticed the couple up here. They thought they could set up a hotel for backpackers. Rhonda's husband liked to travel, so she often had the sunsets to herself.

After looking at Ed's shrivelled chest, Rhonda put on her apron. 'I've got five types of lettuce in my refrigerator,' she told us. Two hours later, we sat down to two plates of beef, at least a quarter of a well-cooked cow, with potatoes, sweetcorn and peas. Ed liberally smeared his with butter and ketchup and was soon halfway through his second steak. I, on the other hand, had forgotten how to eat any meat that wasn't pulverised into a pasty. When Rhonda went to lock the shop, I wrapped my cow in several paper napkins and headed for our bedroom, only to hear the swing door open and Rhonda re-emerge for her keys. I sat on the steak, feeling it seep into my shorts. Addled as I was by days of sun, it seemed the only thing to do. The dogs started sniffing my legs. Rhonda produced more beef, which she skewered on to my plate before sitting down for a natter. 'These dogs are so randy, they must be on heat,' she said politely as one started rubbing itself up and down my leg, salivating. 'Your great-great-grandfather had gumption and he was a Pom. My husband is, too. But he's been reincarnated as an Australian. He used to drive tour coaches; now he lives for the outback,' she said. 'You two seem too young to be tracing your ancestors.'

After twenty minutes of buttock clenching, I lost my grip on the steak and it tumbled to the floor. Ed turned blue. He was choking so hard that Rhonda had to hit him on the back. 'Nasty cough', she said. By the time she had fussed around getting him some water, the steak had disappeared with the dogs. Rhonda hadn't noticed. She was still chatting about her husband, and asked if we'd like to retire to the sofas. I couldn't plonk my gravied bottom on her chintz. Mumbling apologies about feeling faint, I headed for the shower.

By the time I'd sorted out the mess, Ed was well dug in. 'Mount Dare probably looked as green as Ireland when the explorers first came up here.

It can do after the rains. But we haven't had a downpour for seven years,' said Rhonda. She was convinced the rain would be arriving any day now. 'A cyclone's due. If it really rains, the creeks will flood and you'll be stuck here for a month. We'll be spending Christmas together. I've got a huge goose in the freezer. It will be fine. See, the tinsel's already up,' she said, pointing to the mantelpiece. 'We could have a party. There's a nice English girl, just married the property owner up the road, we could invite her too.'

Rhonda explained that people usually moved to this border territory in search of cheap land. 'You can get a whole property for the price of a dog kennel in Sydney. There's no TV, but who wants to hear about wars?' she said. 'My mother-in-law used to send me books, but she doesn't bother anymore. My great-great-grandmother came over on a bride boat from Ireland. The men would stand at the docks to pick the best. I come from tough stock.' Didn't she get lonely? 'Less than in the city. In the town you can go home and run round naked in your living room; here an Aborigine or a tourist is always peering in at the window. I'm very fond of some of the Aborigines, but they don't exactly swap tapestry tips.'

The next morning, we helped her compile the weather reports, which read fine and clear, no rain cloud in sight. 'I get a lot of free advertising for the B&B checking these thermometers. Mount Dare is always the hottest place in Australia, it has its own stick-on sun on the weather map. It's only thirty-four degrees today, but sometimes it's well into the forties.' She easily persuaded us to stay for another evening. A gunmetal sky turned apricot and merged with the hills. As Rhonda's plastic sun disappeared, a purple bruise stretched across the horizon, the desert turned emerald green and the homestead cattle were tinged with red.

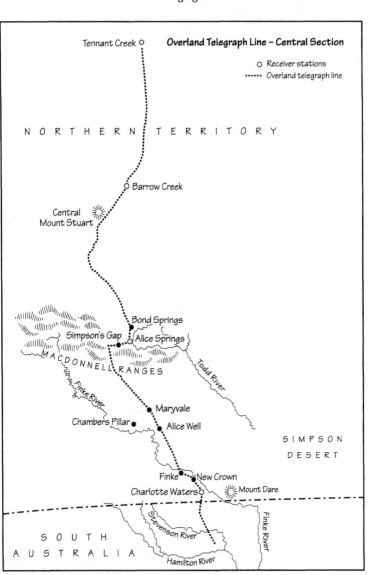

Overland Telegraph Line – Central Section

○ Receiver stations
...... Overland telegraph line

Tennant Creek

NORTHERN TERRITORY

Barrow Creek

Central Mount Stuart

Bond Springs
Simpson's Gap
Alice Springs
MACDONNELL RANGES
Todd River

Finke River

Maryvale
Chambers Pillar
Alice Well

SIMPSON DESERT

Finke
New Crown
Charlotte Waters
Mount Dare

Stevenson River
Finke River

SOUTH AUSTRALIA

Hamilton River

12

· — — — —   ·· — — —

The Rain Tree

We set off at dawn into thick, dappled scrubland, through woods of gnarled trees. A few miles on, we came to a pink signpost, our first for a hundred miles, put there by the indefatigable Pink Roadhouse at Oodnadatta. The sign gave a misleading impression of civilisation. We were 220 miles from Alice, and 500 miles from Adelaide. The road soon petered out, obscured by the drifting sands. A recent isolated shower had made some creeks impassable. We found ways round, and admired the yellow desert plants that sprang from the ground with the slightest aquatic encouragement. Tracing our route with a map and compass, we crawled over the deep ochre sand-dunes of the western edge of the Simpson Desert.

Five hours later we emerged on to a track, part of the Old Andado property, home to Molly Clark. Molly and Mac took up the management of the station in the 1950s. Then Mac was killed in a plane crash on the property and Molly was forced to clear the land entirely of stock for three years as part of a government brucellosis-eradication programme. Molly kept working until a few years ago, when she sold the land and decided to turn her old homestead into a B&B. All meals are cooked on a vintage combustion wood-stove. Ed and I were looking forward to lunch. But

Molly had left a note pinned to her door saying 'Gone shopping'.

Luckily Rhonda had rung ahead to ask the Smiths if they'd have us to stay at their property, New Crown, twenty miles down the road. I opened gate after gate to a procession of home paddocks the size of small towns from which every blade of grass had been stripped. The sand scorched my bare feet. When we drove over some cattle grids we knew we must be close to one of the most isolated properties in the world: the nearest tarmac was two hundred miles west.

We drove through a canopy of gum trees and came to a halt in front of a green, soft lawn. Baskets of geraniums hung from the two houses and the flowerbeds were studded with roses. Feral camels roamed outside the gates, two small boys were rolling between the sprinklers in the garden and a dog was dragging its stomach over the wet grass. A small plane was sticking out of a hangar and several off-road bikes were propped against the walls.

Karen Smith took us into her kitchen, a Goldilocks' paradise with dozens of chairs of all sizes. Her daughter-in-law was making a Christmas cake, another woman was mixing the mincemeat and grandchildren were cutting out star biscuits. They would be fifty for Christmas; all the locals were dropping in, some after a four-hour drive. Ed was handed a beer and told to sit on the porch. 'The men'll be back soon,' Karen explained. 'You'll get under our feet.' I was asked to make some jelly for supper. Embarrassed, I had to admit that my cooking only extended to salads. Karen grinned. 'You feminist townie. You can make the coleslaw, and I'll show you how to use the potato peeler. The men are always starving when they come in from work.' Karen looked thirty in T-shirt and shorts, but already had five grandchildren. Vanessa, the daughter-in-law, was equally pretty.

'How long have you been married?' Vanessa asked.

'Two years', I replied.

'Same as me, only I've got two kids. It takes getting used to doesn't it?' I explained about Ed's bronchitis and she sympathised.

'They're all wimps underneath.'

The men, when they materialised, were all over six-foot tall. Karen's husband, Boof, collected the beers and took them out to the porch for his four sons. In the seven years I'd known Ed I'd barely registered that he had been brought up on a farm. Now his expertise was crucial. While I floundered in the kitchen, failing to be either decorous in my grimy jeans or useful preparing supper, Ed chatted away about milk quotas, mad cow disease and bureaucrats. The men soon forgave his white chest and baggy red shorts as they discussed weather and subsidies, with a light leavening of cricket, the colonial common denominator.

In the kitchen, the women's conversation turned to unsolved murders, the more gruesome the better. As a journalist I wasn't too bad at this game. Soon we'd moved on to discussing the state of British education, and whether girls really pierced their tongues in London. 'Dresses and skirts are much cooler for the outback,' Vanessa advised. I explained that my suitcase full of sundresses was probably in Shanghai. 'Well, there's not much to buy here,' Vanessa said. She'd been teaching the local Aborigines how to sew clothes.

Karen showed me to our room, which had a four-poster bed covered in white linen, embroidered lace cushions, and a dressing table with silver hairbrushes. She worried about a spring clean, but it was so spotless that we were afraid to touch anything for fear of leaving an indelible imprint of the red dust that followed us everywhere.

As we filed into supper, the women sat at one end of the table and the men at the other. The conversation didn't merge until I asked Boof about the property. 'The homestead was a shack when my grandparents arrived. Then they swiped the stone from the old Charlotte Waters telegraph repeater station up the road and lugged it here to build the new house. I got the kitchen put in for my wife, and we're still adding to the place.' He pointed to the stones in the wall above my head which were engraved with telegraph workers' initials.

'There are a few poles left on the property, mostly metal ones, and the last of the wire only came down a year ago. I'll take you mustering if you like and we'll see if we can spot them. The Charlotte Waters ruins are a couple of hours' drive away.'

I explained that I couldn't ride a scrambler.

'We'll use the plane. You'll love it, it's a very effective way of mustering. You climb as high as you dare, cut the engine off and swoop on the cows. Just before you hit the ground, you turn the engine back on and the surprise gets them moving.' Big dippers, corkscrew roads and small planes make me feel sick. Ed volunteered to go instead, but I thought I should try.

The next morning I took three travel motion pills and avoided the peanut butter for breakfast. 'If you're going to throw up, do it into your baseball cap,' Boof said. 'That way we can chuck the whole lot out of the window.' For the first ten minutes I was enchanted. Looking down I could see the outline of the telegraph route running for miles, a faint trace in the scrub, like a vein on a leg. We followed the path over the hills until it fed into Charlotte Waters and threaded its way over the horizon. Several wild donkeys were pottering along the route, cleared to fifteen foot on each side by Todd's men. In this dry country, a hundred years isn't long enough for the vegetation to reclaim the land.

I was daydreaming when Boof suddenly spotted some stray cows, turned off the engine and plunged downwards. Gripping the dashboard, I thought we were about to crash into the line. By now we were almost vertical. I closed my eyes, put my chin on my chest and waited. A shuddering jerk wrenched my head back as the engine spluttered to life and we started soaring up again. This was no eagle's flight. 'The knack is kicking in just before the cows think you're going to squash them,' Boof explained. 'We'll do it a couple more times to get them frightened and the boys can round them up with the scramblers.'

After five rollercoaster rides, I pleaded to go home. Even though it meant a half-hour detour for Boof, my pride had disappeared with my baseball cap.

Boof laughed. 'You're as green as a eucalyptus bush. It's easier when you're actually flying the thing.'

Ed was waiting to go up. 'It's fantastic,' I said. 'See you in a couple of hours.'

It took me an hour before I could face returning to the kitchen, but I was enjoying my new role as woman about the house, and wanted to help make the corned beef for lunch. 'For years this was all Boof would eat, but he'll have the odd potato now,' Karen explained. 'The men used to spend months sleeping out with their horses, checking bores, chasing cattle and barbecuing their steaks along the way. They wouldn't touch a bowl of cornflakes. I'd have to drive out and camp in the bush if I wanted to see Boof.' She looked at Vanessa, who had obviously heard these 'you've never had it so good' stories before. 'My mother-in-law was a real toughie,' Karen said. 'Then we bought the bikes and the planes and now the men have Christmas morning off. No one's allowed to drink until three p.m. so we can play a few party games and enjoy the shrimp cocktails and turkey. But there's not much sitting around.' She sent the children off to sweep the porch and water the plants.

Karen had travelled to England once. 'We saw the castles and went to Shakespeare's Stratford, but there was a queue of cars so we turned round and drove to Scotland instead. A boy who'd worked on the property before going to university invited us to take a rifle and play hide-and-seek with the deer on his estate. I don't know why you don't all freeze to death crawling round after those stags, it's not economical. Planes are far easier. The venison wasn't as good as our meat and those grouse taste bitter. I'm happier here.'

Karen and Boof planned on retiring to the Adelaide hills and breeding Aberdeen Angus. 'You have to get out while the kids are still enthusiastic enough to make a go of it,' Karen acknowledged, 'but we've had a tough few years. They're going to need some luck.' Matt, the second-eldest son, had gone to university in Adelaide when he was eighteen. He came back to help muster during the holidays when the money started drying up with the ground. Vanessa came with him, swapping her parents' suburbs for New Crown. Now they'd moved into the grandparents' old house. Matt occasionally flew up to Alice to play cricket, leaving at dawn and returning shattered at midnight. But no one suggested any holidays. Matt was too preoccupied learning to run the property and Vanessa was busy looking after their two sons, Josh and Cody. 'It's frustrating because it's so traditional here,' she said, 'but if you don't divide the jobs between the men and women, nothing gets done.'

Ed had spent his morning hearing about the drought. We thought it strange, having read various explorers' diaries in which they often talked about pastures of wild orchids, lakes filled with fish, grassy flats and torrential storms, to see the same land leached of vegetation. Boof explained that the early 1870s had been a climatic blip which had only been repeated in the 1970s. Many of the telegraph men had been tempted to build properties in those years when the rivers were full. They'd arrived with their camels, horses and donkeys. Some brought wives and children. The rivers soon dried up. Eventually, they put in windmills to pump water out of the ground, so they no longer had to rely on natural waterholes. Today, the windmills have been replaced by diesel pumps. 'Pumping means we can have far more animals on the land, but when it gets really dry, water isn't enough. There has to be something for the cattle to eat as well. Right now, after seven years with no rain, they've eaten everything we've got. We're thinking of setting up a shoot, luring the Germans over to have a bang at a few feral donkeys,' Boof laughed.

During supper he regaled the table with a story about how he'd nearly been killed when he'd taken a huge bite out of a pie he'd bought at the local Aboriginal store in Finke. Inside it was green and rotting. He is convinced the long-term threat to the property owners' way of life comes from the Aborigines. 'They're only twenty-five per cent of the population in the Northern Territory but they've got fifty-two per cent of the land. Since they got all these free pay-cheques the numbers claiming to be black had gone up from 188,000 to 890,000. Soon they'll be wanting the whole country,' he said, going over to find his map of central Australia. Any land held by Aborigines he'd coloured yellow. 'Before they got the free pay-cheques and the grog, we got on with them well. They were good stockmen. They all spoke English, knew their own customs and respected ours. Now they're dependent on government handouts,' Boof said. 'Given half a chance they come into the kitchen demanding sugar and coffee.' Karen remained silent. But she's the politician of the family, the one who flies up to Darwin to negotiate with the government and who sits on the board of the Isolated Childrens' Parents' Association.

Boof was distracted by the lack of rain. After seven years of drought, the figures didn't tally anymore. Other properties were selling up. The fourth generation of Smiths might have to head back for the shores, he half joked.

Our last view of the homestead stood out as a tableau. Boof was standing by his plane waving, Vanessa held Josh in her arms and was taking down the washing, and Karen was sitting on the gate surrounded by the dogs and green grass.

We worried about the Smiths intermittently. Then, a year after we visited, Vanessa wrote us a letter. 'We have finally had good rains and the feed is abundant. It has created a nice living environment for the gardens, not to mention the fat cattle that are now roaming the desert! I've acquired a pet camel calf called Clementine who attacks Karen's washing. The children went to the Kulgera gymkhana and Joshua's started Alice Springs school of the air.' The property looked as though it would feed another generation.

Her next letter was far sadder. Greg, Karen's and Boof's youngest son, had died in a car crash. But his death showed the strange intertwining of the outback's two communities and how reliant they really were on each other. 'The response from the Aboriginal people in Finke was very touching,' Vanessa wrote. 'They came to our family service for Greg at the station, where we put some of his ashes under two rose bushes. They brought with them a rain tree which they planted near the roses in our garden and as they filled in the hole there was complete silence in respect for Greg's memory. Forming a long line, one by one they offered their condolences by either a handshake or a hug, old men and women and young. It was the most spiritual experience I have encountered in my life.'

*

We had to drive through the town and ford the dry bed of the Finke, the oldest river in the world, to head on up towards Alice. Finke, once the second depot on the overland line, now has a smattering of houses lived in by Aborigines. We couldn't find anyone to sell us petrol. Adults hung around looking bored while children played with dogs in the heat. The shop sold oozing cheese and rotting potatoes, but strictly no alcohol. The Aborigines appeared to live off the same diet as us: crisps, biscuits and ice cream. A white man had been employed by the community to look after the shop and carry out repairs on the houses. A white teacher came up in term-time to run the school. The Aborigines had nothing to do except drive through the bush.

One of their cars, a jeep, was bogged in the sand of the dried-out river and the Aborigines came to ask us for a tow. While they played our tapes and drank our Coke, we attached an old rope we'd discovered in the boot of our Toyota. It snapped. We found a tar-stained black metal coil under a spare tyre and tried again. Ed pushed the jeep as I drove. Only a one-legged grandfather volunteered to help. At the Smiths' we'd felt uncomfortable with their harsh talk about the Aborigines. I'd read dozens of books before arriving in Australia about the shoddy way the indigenous population had been treated. As a teenager, I'd been mesmerised by *Walkabout*, a film about a white teenage girl who is led through the centre by a young Aborigine. I knew more Aboriginal myths than Australian bush-songs. But it was hard to stay romantic while covered in oil and exhausted from digging out their jeep, as five men sat under a tree waiting for us to solve their problem. There was no chance of a shower before Alice. Within seconds of our pulling them out, they'd bogged the car again and we had to repeat the whole process.

Depressed, we almost missed a set of telegraph poles disguised as goalposts on the football pitch just outside town. There were four of them, all metal, as white and straight as the flagpost on Buckingham Palace. From here we had two options: either we could drive on top of the old Ghan line or we could roller-coaster along the parallel hummocks of sand keeping up a steady twenty mph, so we didn't get stuck in a rut. For an hour we veered between the two, both excruciatingly uncomfortable, before arriving at the turn-off for Alice Well. This was the waterhole discovered by Ross which became a main depot for the central teams.

A tree, scarred with initials, marked the path to the well, now long overgrown. The temperature was in the forties, and we couldn't face a two-hour walk into the bush, so we risked taking the Toyota off the main track: we made it to the well in time for lunch. A few palm trees were scattered about. It looked quite biblical from the car, although the bore-hole was now covered up. Too hot to scavenge far for remains, we found a few beer bottles and an old shirt, but no poles. There weren't even the mosquitoes the men had complained about so bitterly, only two wedge-tailed eagles in search of food. We turned the car round.

That was when it started spitting sand, burying itself further and further into the ground like the emus we hadn't yet seen. After a few moments it was clear we were stuck. Nobody ever came this way at this time of year. The track along the Old Ghan line was disused, and we were well off it anyway. There were no cattle stations for over a hundred miles. We hadn't yet grasped how to use the radio. There was, we told each other, no need to panic.

Remembering Todd's instructions, we played scrabble until it got cooler. Then we started digging. After an hour the sand had embedded itself in our tar-smeared bodies, our hair was plastered with sweat, I'd lost one of my contact lenses and the skin on my neck was bubbling. 'Darling, you look fantastic,' Ed said. He tried looping a chain around a tree, wrapping the end round his waist and getting me to rev the car. It didn't budge. The electric winch, which we had refused to take on the grounds of overkill, looked like a belatedly sensible option. The wedge-tailed eagles started pecking viciously at our discarded picnic. In desperation we took two loose planks from the bore, put them under the front wheels and Ed revved us out. We skidded back on to the railway line, laughing maniacally.

If we wanted to be accepted in Alice Springs, we were going to have to find some water to clean ourselves up. Looking at the map, we found that the only chance was a hundred miles away at Maryvale Station. Having crashed out, exhausted, under the initialled tree, we woke at five a.m. the next morning. I was soon panicking that my sunburn was going to give me skin cancer. By seven-thirty we had reached the end of the old railway line and joined a property track. For all its dust and corrugations, this was civilisation.

By eight a.m. we were digging out three Aborigines who'd spent the night stuck in a ditch. The Aboriginal community lay just beyond Maryvale Station. Every cattle gate leading to the homestead had been left open, bottles were strewn along the wayside, discarded by the Aborigines before they reached their 'dry' community. Signposts had been turned around or taken down. Several cars had been abandoned, their doors open, slewed across the road.

We got to Maryvale by ten. The store was closed, but a woman came out of the homestead and we explained our predicament. 'All you need is some eucalyptus oil; that'll get the tar off immediately,' she said. I explained my allergy. 'Well, it's that or staying tarred.' She took us round to her backyard and showed us the hose. We stripped to our underwear, smeared our bodies with eucalyptus oil and started rubbing. Ed hosed me down just as a rash started appearing. We cut the bits of tar out of our hair, changed into clean shorts and decided we looked respectable. 'How long have you been out in the bush?' the woman asked, as we rifled through the shop inspecting the sell-by dates on the fruit cake, stocking up on cans of sweetcorn and packets of ginger biscuits. There was no milk, so we ate three ice creams for breakfast.

The woman had moved in six weeks ago, taking over the property from the Hayes family, who, we had been told, were the true aristocrats of the outback. They had finally buckled under the drought and sold their smallest station to a conglomerate in Sydney. 'At the moment, the only money we're making is from selling Christmas baubles to the Aborigines in the shop. It's funny: you hear all about their ancestors' spirits and how important certain sites are to their culture. But they think they invented Christmas as well,' the woman said. 'One old man told me the reason the Hayes had been forced to sell this station was because they'd massacred all the donkeys. He said they'd been cursed for killing the Aborigines' holy animals. But the donkeys only came over here with the white man, didn't they? They've got their dreamings and their Bible muddled up.'

We didn't want to get back to shopping malls yet, so we thought we'd camp out at Chambers Pillar for one last night before hitting town. The track to the valley descended the steepest gradient we had seen for weeks. As we set up our camp at the foot of the pillar we saw our first kangaroos, five of them bounding in perfect time towards Castle Hill, a red stone outcrop resembling Windsor Castle. We examined the pillars for Giles's initials, but they had been obliterated by an autograph album of later explorers.

Central Australia's landscape is pale cream and red sandstone, remnants from a Cretaceous ocean floor. Mostly it is flat, which makes Chambers Pillar, and the surrounding rocky castles, all the more surprising. The early explorers used it as a guide. Now, it is an isolated beauty spot, stark and stunning in the evening sun. Aborigines call it Itirkawara, and say it is the transformed fossil of a knob-tailed gecko, a sandhill lizard. This ancestral gecko was an aggressive warrior who travelled across the desert, killing men and capturing women. He was so evil he fraternised with women from the wrong kin-group. Eventually he brought a young female captive from another group to the camp of his own family. They were outraged and exiled him. As he led the girl across the dunes they became weary and rested. Itirkawara's male organ turned into this luminous pillar of sand. Nearby, Castle Hill is Itirkawara's wife, crouched down in shame and averting her face. Stuart called it 'a number of old castles in ruins'. But like the biblical story of Lot's wife, Itirkawara is the Aborigines' warning of the importance of observing strict social codes. In the sunset, the castle flared up like an awkward blushing bride. Every time we woke during the night we'd see the pillar illuminated by the moon.

13

· —— —— —— ·· · ——

A Town Like Alice

The next morning we drove through Heavitree Gap, such a perfect clear break in the ranges that it seems impossible that Ross could have missed it. Then we passed the railway station and came to our first roundabout in five hundred miles. By ten-thirty a.m. we were in the outskirts of Alice. There were signs to squash courts, libraries and beauticians. We saw the first pavements for five weeks and telephone poles ran along every street. Law courts, the town gaol, supermarkets and an Overlander Steakhouse all jostled for position.

The heart of the town is the open-air Todd Street shopping mall. It was filled with clean-cut Japanese and American students ordering croissants and cappuccinos for breakfast and nursing blisters from their new thongs. I decided my ancestor would have enjoyed giving his name to this symbol of mercantile civic virtues. Both Todd and Alice have been awarded plazas. Todd's boasts the Jolly Swagman pub, Alice's plaza is home to Fawlty's and Doctor Lunch. She also has the Old Alice Inn, while her husband has Todd's Tavern, with steaks at A$8 and a salad bar.

We'd decided to stay at the Diplomat, nicknamed 'Miss Daisy's' and the oldest hotel in town. The original building had burnt down and now it

· 118 ·

looked as new as its neighbours, with plastic grass round the swimming pool. In its café, local businessmen drank Sauvignon over deals and shouted above the sound of traffic. There were slippery corn-blue nylon bedspreads in our room, a jacuzzi for a bath, an ironing board down the corridor and cleaners gossiping by the Coke dispensers. All rooms overlooked the parking lot, but were framed by purple begonias. The place was surprisingly jolly. The receptionist handed us new toothbrushes and razors, and pointed us to the laundry room, providing a fistful of coins for the laundromat.

Miss Daisy's used to cater for bushmen; now it's a staging post for Ayers Rock, officially known by its Aboriginal name as Uluru, a more glamorous tourist site than Alice. When the Rock managed to arrange enough irrigation for a golf course and a few swimming pools, many big hotel chains moved there. Fewer tourists are spending a night in Alice. They can buy their stuffed kangaroos and boomerangs at the airport and catch the tour buses straight to the red rock. Alice is gradually reverting to the polite small town of Nevil Shute's book, A Town Like Alice – only instead of ice-cream sodas, it is full of craft centres, tofu salads and bookshops. Plans to build a longer runway at the Rock could turn Alice, currently the only central stopping point for big jets, back into the quintessential outback town.

There was no sealed road to Alice until the army built one during the Second World War, and the fully sealed highway was finished only in 1987. Frank Wright was one of the first men to cross Australia in a car. In 1929, he took a rugged Vauxhall nicknamed 'Vauxie' through the centre. On reaching Alice, he wrote: 'The town is nothing to write home about; a pub, a store, a Residency and a few houses.' My sister was equally dismissive about her teenage visit with a backpack in the 1980s. 'It was a relief to get off the bus and go to the loo.' But I'd already grown attached to Alice. Set against the blue skies and red rocks, the Acacia Roads and Orchid Avenues seemed surreal rather than suburban.

The town has managed to entice the highest density of scientists in Australia, many working on mining projects. It also boasts a 'secret' American base nearby, which provides gossip for the town as well as monitoring the airwaves on behalf of the world's superpower. Property owners visit to stock up on food and Aborigines come in from their communities to buy liquor at any one of sixty-eight grog outlets. It was hard to see what the majority of the population was doing here. Our hotel manager could have been heading any Four Seasons Hotel. He was the perfect diplomat, with a well-groomed handlebar moustache and tight, pressed shorts, but he preferred it at Miss Daisy's. He couldn't explain what kept him in Alice except habit and the lure of being in the centre of several thousand miles of desert. We speculated that it was the moral as well as the physical space. For a small town, 'the Alice' is remarkably broadminded. And, unlike other small towns, it accepts you as a local after a week.

We met another resident, Jose, in a newsagent. Wearing a tea-dress and hat, the octogenarian was signing copies of her new book on the origins of Alice's street names. She invited us back for tea, insisting on making us some bachelor's buttons, 'not that bought rubbish'. But she compromised on the teabags. 'You can't get loose tea anymore in these supermarkets.' While she rolled the biscuit mix, Jose reminisced about how she had come to Australia as a governess from Southampton. Between the two wars she had met Hans in Cairns and followed him up the dirt track to Alice. Hans had worked in the sugar-cane fields in Cairns, but had damaged his back and decided to chance his luck on the cattle stations instead. Jose preferred the dry heat to the humidity of Queensland. After his death in a freak accident, Jose retired to town and introduced her neighbours to dog-training classes.

Her son drove one of the huge refrigerated road-trains along the Stuart Highway, bringing food to the centre. 'They're the only real men left now,' she said. 'They're so tough, they'll do the coast-to-coast trip in three days.' Her grandchildren were now based in Adelaide and the last signature in the visitors' book was six years ago. She had stopped teaching the Girl Guides because the young weren't interested in winning badges anymore. Her friends used to enjoy annoying the American airbase by having picnics just outside their grounds and playing the bagpipes. But the Americans didn't rise to the bait these days. She would never leave Alice Springs. 'I'd be bored to death in Southampton. I'm a Centralian,' she said. It was the first time we'd heard the word. It seemed clumsy but apposite; the Centralians versus the Coasties.

The next day, a letter arrived at Miss Daisy's. It was from Jose, pleased we had eaten twenty of her biscuits. The note read: 'Here's the recipe for the Bachelor's Buttons – look after "Ed"! Is he your editor?' There was a book enclosed, called *The Art of Living in Australia*. On the first page there was an inscription: 'To Prudence, with all my love, 1895'. It gave Victorian tips on how to clean one's teeth in a hot country, how to bring up children and, most importantly, a series of recipes for Australian life. I looked up our biscuits:

Bachelor's Buttons

½oz flour
1 egg
2oz sugar
½ teaspoonful carbonate of soda
1oz butter
1 teaspoonful cream of tartar
6 drops of essence of almonds

Rub the butter into the flour, stir in the sugar, soda and tartar, mix into a stiff dough with the egg and flavouring. Roll into small balls the size of marbles, toss in coarse sugar and bake from five to eight minutes.

There were tapioca meringues, hasty puddings, Stanley puddings and Yorkshire teacakes. The book was written by a Doctor E. Muskett. I spent the afternoon reading about how to cook your husband a healthy breakfast, not sparing the butter, and why one shouldn't be snobby about the 'Brussel's sprout. A salad is a delicacy which the poorest of us ought always to command.' Food in nineteenth-century Australia was slowly improving, and these were the kinds of dishes the Todd family would have eaten every day. I decided that when I returned to Britain I would master the stuffed flathead, oysters and bacon, and tripe in milk.

Ed dragged me to see *101 Dalmatians* at the only cinema after making the obligatory stop at a doctor's to make sure his chest was recovering. It was packed with rednecks in search of air-conditioning. For the next couple of days, we ate spaghetti with clams, Caesar salads and tried the local Château Hornsby wine, until we became bored with the novelty and returned to what Alice does best: steak and sliced white bread. We lay under our fan, mooched around the mall and along the riverbed, inspected our moles for the first signs of skin cancer and made every excuse not to clamber back into the sticky car seats.

By now we were ready for a small excursion. The Springs are located two miles outside Alice, alongside the telegraph station. When the new town had been built one step south in this more sheltered valley, the planners had changed the name to the more macho Stuart, after the explorer who had blazed the trail for Todd. The locals hated this meddling and Alice soon reclaimed her metropolis. On a Sunday we joined a procession of families making their way along the dry Todd river for a barbecue on the spacious water-sprinkled lawns of the Springs. The lunch-time heat was so strong my fingers swelled to double their size and the Springs themselves were a puddle, but the breeze cooled us down as we ate choc-ices under the ghost gums and watched the red kites flitting through the shadows. No wonder Todd loved the place, with its water and shade.

The original telegraph station has been renovated with fences and sign-posts and is run by the local park rangers. Muscular guides, dressed in Victorian pinafores and laced black boots, were showing sweating Australians round their heritage. Every hour they gave a speech about Telegraph Todd and the overland line, asking children to guess the number of poles needed to reach across the continent. The rangers added graphic descriptions of the bleak life on the isolated telegraph stations and anecdotes about explorers having to sleep in their saddles for fear of snakes.

The Australians were attentive. For them, this was a shrine to the outback myth, the emotional heart of their history. In Britain, many stereotype Australia as the home of Bondi Beach, *Neighbours* and newspaper magnates, but urban Australians see their country's identity in terms of billabong tea, campfires and damper. They queued up to buy pencils and beer-coolers stamped with 'I've been in Alice.' Ed thought this funnier than I did.

A small museum showed a jumble of early settlers' memorabilia. In one corner was a silhouette of a family of eleven children, with a young girl sitting in the front. It was Alice. Evidently the Queen had been sent the silhouette by one of my more gushing British relatives and had given it to the town when she was passing through in 1963. The Alice residents thought the idea of the Queen packing a silhouette next to her underwear quite normal. Their British relatives always brought presents and, until the politicians in Canberra decreed otherwise, the outback felt they enjoyed a personal relationship with the Windsors. Alice was the place where Diana, Princess of Wales, outshone her husband for the first time. Monarchist shopkeepers asked after William and Harry, Beatrice and Eugenie. But the Queen's role as Commonwealth curio courier was new to me.

In another corner was a single photograph of a group of half-Aboriginal children, with a short description mentioning that they had lived by the springs from 1932 to 1945. The guides passed them by. The history of these mixed-race children, who had been taken from their parents and brought up in orphanages, was not for a family day out. One of the guides quietly admitted that the policy had been a disaster. The state had managed to alienate the children from both sets of roots. After the Japanese bombing of Darwin in 1945, Alice Springs station was turned into a Native Labor Corps Headquarters. Over two hundred Aborigines were kept here and were delivered to the nearby army base every morning by truck. Vic Hall, the sergeant in charge for a period, praised their keenness and good work but found them 'being fed a diet unfit for anybody . . . while their women and children starved'.

As we wandered back into town, we watched the groups of Aborigines along the dry Todd riverbed, surrounded by lottery tickets and cans of beer.

In Todd Mall, they sat lethargically on the litterbins. Most of white Alice reacted in the same way whenever we mentioned the subject. First they would enthuse about the dreaming stories and extol the virtues of the Aboriginal art in the galleries, saying they were a matter of reverence and national pride, like a Turner or Gainsborough. Then they would admit that white policies over the past two centuries had destroyed the Aborigines' independence. But get them on to the present and they would open up. A small reservoir of embarrassment turned into a torrent of frustration. The two groups seemed to be ruining each others' lives.

What was Todd's part in this? According to Pat, Alice had been known for her kindness to the Aborigines. When they arrived in Adelaide every May to get their winter government ration of a blue blanket, she had supplied them with beef and the children with milk. Lorna recounts meeting a group of Aborigines when walking with her nurse one day. 'I had inherited from my mother and father a liking for these dusky-faced, white-teethed men. They had such kindly eyes. I was not surprised when they stopped me. "I know you, you Telegraph Todd's piccaninny," they said.' Lorna describes how they camped on the parklands just outside the high paling fence that ran around the observatory grounds in Adelaide.

> Every morning they would come to the Observatory kitchen with their billies and line up for a ration of tea and sugar, bread and meat, for they were our friends and our guests . . . They often staged a 'corroboree' dance for us. With us sitting on the sharp edge of the fence, to watch the warriors with their bones outlined in white. If Telegraph Todd was to be in the party, they put on a special show. The whole family watched their procession as they left town. They went in single file, warriors in front, the lubras following, household possessions on their backs, with a bright eyed piccaninny often peeping out from the top. 'Where are you off to?' we would cry. 'Going longa Goolwa,' was the inevitable reply, so we supposed their destination was Goolwa, Port

Elliot or Port Victor to the south of Adelaide where they had their camps.

Lorna remained adamant that Todd was enlightened compared with his contemporaries.

My father made many friends among the blacks while laying the lines and he used them whenever possible to help. Some years ago, a man riding on a journey along one line noticed a black boy climbing the poles and testing the insulators. When he got down, he tapped the pole three times. 'Why do you tap the poles?' the man asked. 'Tell Telegraph Todd done 'im,' the boy answered.

Yet Todd had blood on his hands after one attack on Barrow Creek telegraph station. One summer night in 1874, five station-hands were sitting chatting on the south-west side of their courtyard. Suddenly a volley of spears hailed down on them. As they ran round the building, they found their way blocked by a group of natives. One man, called Franks, was speared. He was just able to reach the kitchen before he died. The temporary stationmaster, James Stapleton, was trying to close the gate when he was hit by four spears. The other men dragged him inside but Stapleton was dying. So they went to the key and tapped out details of the attack down the line. At the GPO, Todd was alerted and immediately sent his carriage to bring in Stapleton's wife and a doctor. Stapleton, a pioneer of Canadian and American telegraph systems, asked the men to lift him to the keys. Todd wrote down each letter as it arrived and handed the slip of paper over to his wife. It read, 'God bless you and the children.' Stapleton died the next day.

Todd organised a concert at the Observatory to raise money for the Stapleton family. But he turned a blind eye to a punitive expedition against the blacks by a party of police troopers and telegraph men. Although official reports claim there were no arrests made, they fail to say why. The search group scoured the area for months and killed every Aborigine it could find. The police officer in charge merely wrote: 'The natives never had such a lesson.' So I had my own residual guilt, which made me more Australian than my Akubra hat.

When I returned to Adelaide I went to the South Australian museum, partly for research and partly because everyone had told me that the best-looking man in Australia ran the Aboriginal department. Lesbian acquaintances admitted they'd jump ship for him. 'He's utterly divine,' they said. Philip Jones was six foot four, brown-skinned, blond-haired with a distracted Indiana Jones air. Hollywood handsome, he was only just beginning to realise it at thirty-five. Philip explained that he had always wanted to be an anthropologist, despite the fact that his relatives ran a successful breadcrumb factory. The family bakers had moved from Eastern Europe to Adelaide in the mid-nineteenth century. He had spent the past ten years writing his thesis, 'A Box of Native Things', rummaging around in the outback for Aboriginal remnants. In his office he had a full scale model of Daisy Bates, the Florence Nightingale of the Aborigines, a few bones and his computer.

He confirmed that the telegraph route was the beginning of the end for the nomadic way of life. 'That's when they began to modify their behaviour. The Aranda tribe was the hardest hit. They had to share waterholes with whites and learnt to barter,' he said. 'Todd would have been suspicious of the blacks. A man on Goyder's expedition to the Northern Territory had been killed the year before. He'd got too involved, took part in ceremonies, ate turkey with them, hunted with spears and bathed in their lagoons.' That would be why Todd told the overseers to keep fraternisation to a minimum? 'Exactly', Philip said. He took me down to the storerooms. There were rows of drawers, filled with hair, glass tips and spears. From one chest he took out Todd's collection of Aboriginal implements, four knives and a tomahawk, all with Todd's name attached to them on yellow tickets. Todd's Aboriginal spear was on tour with an exhibition.

'But I thought Todd wouldn't allow the men to collect any artefacts along the line,' I said.

'He was the boss, wasn't he?' Philip smiled.

Todd would offer insulators as tips for spears, he bartered shards of glass and metal in return for coral necklaces for Alice and amusing curios for his study. 'The noble savage' had become a major attraction at international exhibitions. Scientists collected weapons as they once collected fossils. 'There is a theory that the Aborigines were on their last legs anyway. The Edwardians thought they were simply primitive relics from remote times and their brains would never develop,' Philip said. 'They were a curiosity merely because they hinted at our Stone Age ancestors' way of life. The white man's burden was just smoothing the dying pillow. But that's rubbish. The Aboriginal way of life was highly developed. The line must have astounded the tribes. Their own systems of information transmission, using messengers, ceremonial gifts and smoke indicators, were so much more social. But they were drawn in by the labour-saving food, sugar and salt that they found at the camps, depots and repeater stations, to which they created shanty annexes. The telegraph stations became ration depots, pasturalists ruined the land, ancient flora disappeared under cattle's hooves.'

Philip said it was impossible to assess the impact that the graffiti on rocks like Chambers Pillar had had on the morale of the Aboriginal tribes. But it was clear that the desecration of their dreaming rocks caused huge resentment, as well as the white man's use of the watering holes, and may have been the cause of many of their 'unprovoked attacks'.

We drove out to the rangers' station to search for some of Todd's own relics. The park ranger on duty looked nervous when we explained the family connection, realising that we would want to see Todd's memorabilia. Pat had given the station his awards, medals and scientific instruments and other relatives had added to the hoard. The rangers had tried displaying them at the telegraph station, but the building had been broken into and they'd lost Todd's revolver. They couldn't think of any other use for them, so they stuck them in a shed along with a collection of three hundred preserved animal skins. The skins had been donated by the family of one of the last great hunters, who had collected examples of more than forty species of indigenous animals at the turn of the century. The rangers were terrified that we were going to complain, but the medals looked at home with the carcasses of now-extinct beasts flapping in the breeze from the open door. The barometer, whisky flask and powder keg appeared in pristine condition. At least here we could touch them.

Over a beer, the rangers explained that this was one of their favourite postings, in charge of hundreds of miles of scrub and miles away from the bureaucrats. Their only problem was the recent discovery that Alice was drowning in too much water. The gardens and swimming pools now used so much of the stuff, piped in from aquifers in the desert, that the water table under the town had risen to alarming levels and was threatening to drown the roots of trees.

One guide was fascinated by Alice. She wanted to know whether she'd made the trip up to her namesake. But, despite the tale about her piano arriving on camel-back, my great-great-grandmother never travelled more than thirty miles from Adelaide after her disappointing trip back to Cambridge. She'd once written to Todd, while he was away supervising the overland line, begging him to take her on a holiday when he finished his great work. But they never made it. When the elderly Todd went on a tour round the great European capitals or attended inter-colonial telegraph conferences, he took a daughter for company. Alice was needed to run the family home. Yet he always swore he was so devoted to his wife he could hardly bear being parted from her.

Feeling guilty that I had dragged Ed through the equivalent of two months of old family photograph albums, I asked him what he wanted for Christmas. 'A holiday,' he replied. So leaving our car and dirty washing at Daisy's motel, we booked a flight for that evening. We were going to an island off Queensland where the sun was cool enough to get a tan, the air was humid and we'd be surrounded by water. We could scuba dive in the morning, water-ski all afternoon, play tennis and go for picnics on deserted beaches with baskets of lobster and fruit. It was inconceivable to think of Todd in a wetsuit, so I promised not to mention either Charles or Alice during our break.

We evacuated on Christmas Eve with half the American airbase and arrived back in Alice to celebrate New Year with a cheese fondue in forty-one-degree heat. Within minutes of returning from the Queensland coast, I couldn't help blurting out that I sided with Todd. 'What about?' Ed asked. Bringing a telegraph line through that tropical jungle would be madness, I explained. The rain would bring down the wire faster than a falling coconut. The tree trunks were too wide to make poles and the dry red sand is merely substituted for wet yellow sand.

For six months after South Australia started its grand project, Queensland continued to write to London, complaining that their rivals were bound to fail. But if I'd had to choose, I'd have gone through the centre.

14

·— — — — · · · ·—

The Women's Tale

The Alice tourist industry has tried to think up gimmicks to entice foreigners to stay in their town. In July, they hold the Camel Cup and in September the Henley-on-Todd regatta, so that the locals can dress up in their yachting blazers and run down the dry river holding their bottomless boats. In the summer they promote balloon trips and camel rides. The latest ruse is bush-walking. Todd would have been envious at how easy it is to pick up guides to take you around the two hundred miles of the MacDonnell Ranges. When he arrived at Alice Springs to inspect the new depot, Mills and McMinn had only recently found their way through Simpson's Gap.

Ian, a stocky ex-British army officer, thought we were mad when we asked him to take us on a five-day safari in the glare of the summer sun, so we could experience for ourselves how difficult it was to find the hole through the MacDonnells. Originally from Zimbabwe, Ian had decided to set up a wildlife trekking company in the dry heat of Australia while crawling through the swamps of Borneo as a jungle-warfare training instructor. His company, Trek Gondwana, taught city-dwellers about the kinder side of nature. Australia was not Africa, but it had enough similarities and it wasn't likely to have a civil war, he'd reasoned. When he'd first arrived, Alice Springs had reminded him of old Rhodesia. He agreed that desert would be preferable to the tropics when it came to planting telegraph poles.

Ian tried to recreate the lifestyle of Ross and Giles, but we were allowed baked beans with our bully beef. He told us which tree bark was good as an antiseptic, which roots produced moisture, which snakes were poisonous and which frogs' saliva made you hallucinate. The former officer was a great deal better trained than any of Todd's men. 'Surviving on goannas is easy. In Borneo we only ate grubs one week and if we wanted to defecate we had to carry it with us in plastic bags so no one could track us,' he explained. But they all had radio contact and if a soldier went down with malaria at least he could be airlifted out.

Ian thought the telegraph men had been badly briefed, despite Todd's manual. 'They didn't bother to learn the basics off the natives,' he said. But then Ian had had the benefit of a Sandhurst officer's training. We had to explain to him that South Australia, in the nineteenth century, didn't attract officers. They were off playing the Great Game in India before returning in triumph to their villas in Sussex. South Australia's telegraph pioneers were craftsmen and religious dissenters. They hadn't been trained to fight for the Empire; they were trying to create their own mini-Utopia.

We dived in icy springs that the telegraph men never found, and walked among the cycad palms searching for wild turkey. Ian was horrified when I suggested baking the turkey in mud. He was a conservationist now, so it was all corn-flakes and Carnation milk in biodegradable packages, and lettuce and Vegemite sandwiches.

Every night, Ian would point out the southern hemisphere stars that Todd had helped map, and at lunch we'd sit under silvery ghost gums. If you squint your eyes, said Ed, the purple mulga could almost be Scottish heather. Above us was the ridge of the MacDonnells, a green escarpment which rose sharply from the flat, arid land to the south.

The range looked impenetrable, but as we came closer we saw a thin sliver that at first looked like a tall tree trunk. We trekked through dense sand until we came to the entrance to Temple Bar, a crack just ten foot wide guarded by thick vegetation and a freezing deep pool of clear water. We stripped off and swam to the far end, winded by the cold. Pool after pool stretched deep into the range, where the sun never reached. Having left our clothes on the south side, we decided to return and find another route.

Further west, we came to Simpson's Gap, where Roe Creek has gouged a red gorge through the Rungutjirba Ridge. Only now it had been renamed Honeymoon Gap and as we emerged from our scrub we saw it was crawling with tourists who had driven to it along a tarmac road. Buses queued up outside the gates to disgorge passengers, who traipsed back and forth along a wooden path between the refreshment kiosk and the gap. Some tourists seemed disappointed, but there was an entertaining group of rock wallabies playing on the hillside. At their feet, a thin layer of suntan oil covered the surface of a murky pool of water.

After five days without people, the line of fluorescent shorts and portaloos was too much for us and we retreated back into the bush. Ian thought this was why everyone would eventually come to Alice. There was so little space in the world and she had so much of it. He had once been asked by a Hollywood mogul to find a cave three hundred miles from civilisation. The search took a month before Ian found the perfect spot. The mogul then got him to scrape out the bird droppings and helicopter him and his nubile girlfriend to the cave with a week's supply of food and drink, a stove, some bearskin rugs and his cigars. As the helicopter landed, he handed Ian his mobile phone, threw his bikini-clad girlfriend over his shoulder and disappeared.

At the other end of the tourist scale, we witnessed a $10 night-walk. Ten Aussies joined us at our campsite for cold beer and enjoyed berating us for our flat, Pommy stomachs and preference for wine. After the coleslaw and steaks came the highlight of the evening. Ian gave us a safety lecture on night-time safaris. He'd introduced them from Africa, where tourists search for lion and leopard using jeeps and spotlights. This one was to be on foot. We must be vigilant and silent at all times and take the hand of the person in front, he told us. We set off into the pitch-black bush, stumbling over rocks and smashing through spinifex, with the men all swearing as they trod on their wives' feet. After an hour of seeing nothing, Ian must have sensed that his troops were getting restive. Suddenly he stumbled on a large beetle and we all gathered round to examine the phenomenon. The group was entranced. A few kangaroo droppings and they were happy to return to Alice. The next night they were going to learn how to sing camp songs.

Ian offered to take us to the other side of town, a fifteen-mile dust drive away, to visit the Eyres family. Bill Eyres the First had come up to Alice a couple of years after Todd, to start replacing the wooden poles. But while Todd had decided to consolidate his career in Adelaide, Eyres had stayed to start a property. Now his great-grandson was trying to keep together a place the size of Belgium for his sons, Bill and Fred. It could so easily have been my father farming in this heat haze. Bill Eyres Senior offered to meet us at his new visitor centre. He had decided he must move with the times and had built a restaurant, shop and sunset-viewing platform on his land. The bulldozers were still scraping topsoil away as we arrived. The station workers turned up with a young apprentice who had saddle sores from bumping up and down in the middle of the Land Rover.

Bill had been to Sydney once to find a wife, but hated leaving his property. Alice was his lifeline for new computer disks, loo roll, aeroplane spares and first-aid equipment. The Sydney lass he had brought back forty years earlier was now a true outback lady. At nearly six foot tall and two feet wide, and dressed in a green and yellow elasticated tent, she revelled in shocking us with tales of rough living in the Centre and stories of how she used to ride for four days to see a girlfriend. On her wedding list she had put down toaster, iron and hairdryer. But it was the 1950s and there was no electricity on the station, so her presents lay unwrapped under her bed. Later they installed a generator. When the men were off gathering the cattle, she would be left to walk in her nightdress the half mile to turn the electricity off. She still cooked a traditional lunch every day, serving plates of boiled beef, carrots and boiled buttered potatoes, followed by apple-and-peach pie. The only modern concession was ciabatta bread, which a fashion photographer had shown her how to make. It transpired that *Vogue* occasionally came out to the property to shoot against the hills. Mrs Eyres had never seen the magazine but she had met the supermodels Naomi, Kate and Elle, and had been told she appeared in the background in her apron in one shot. 'The men here didn't go for them. Not enough belly,' she laughed.

Bill teased me about my family's return to Britain. 'You can't call yourself outback aristocracy,' he said, and showed us pictures of his ancestor, who would have towered above Todd in his white breeches. Mr Eyres was proud of his family's intimate involvement in Australian history. He said it still gave him pleasure stumbling across the original wooden poles that hadn't been knocked down by the cattle scratching their backs, and that he kept some of the insulators in his study. He even drove us to see some metal Oppenheimer poles still standing like masts of shipwrecks scattered over the hills.

For all his acres, Eyres, like the Smiths, was in financial trouble. The new planes and bikes cost a fortune. He also had an Aboriginal camp at the end of his land: 'Didn't you see all the beer cans? Kills the cattles' hooves.' He liked the 'abos' but he'd lynch a liberal white lawyer if he found one on his land. Where once he had let the Aborigines walk freely over his property, now he chased them off. He was terrified they would 'discover' some ancient ceremonial site and claim their 'rights'. It had happened to other property owners. Lawyers were the only winners. Life in the Red Centre was hard enough without this state of perpetual legal siege.

We drove on to look for the Pinch, a smooth, chiselled path between the Ooraminna Hills which had been hacked out by the Afghan cameleers and was used by all traffic to Alice for half a century. One traveller wrote: 'In reality, it forms a kind of smooth rock ladder, up which the horses stumble as best they can.' It looked like a giant bobsleigh run, with the initials of successful competitors carved into the side-walls. That night the thought of traffic seemed remote, so we decided to camp in the middle of the track, forgetting that there was a bore-hole up ahead. At four-thirty a.m.

the first cattle started picking their way over our swags. We soon had to scramble out of the way.

Later, after Ian's fried eggs, we went in search of some rock caves. The country seemed much older than the MacDonnells. It was worn where they were jagged; red where they were green; barren where they were fertile. This outpost used to be twelve days from Oodnadatta. Now we were only an hour's drive from Alice. Ian had to drop some leaflets off at Bond Springs, once a repeater station to the north of town. 'Ah, you're the telegraph girl,' the owner said. 'We had the contract to pull that thing down thirty years ago. We used some of the poles around the station. I think a man named Bond was on the first trip.' There were even poles as washing lines.

Diane, Ian's wife, came to meet us in a café in Todd Mall with her friend Jo. They were discussing how to keep their furniture from warping in the heat. Jo had come out from a British boarding school aged eighteen for a year on a property and met Ben. Five years later she came back to reclaim him when he started a job with Ansett airlines. The heat hadn't damaged her English skin or diminished her passion for chocolate. 'It wasn't such a risk marrying Ben. You can make a home anywhere,' she said. 'A bit of Pommy pussy's what's needed. God, I'm getting crude out here. In London I went to dinner parties every night; here we just party at weekends. But they're real parties.'

Centralian women aren't crude, they're feisty. Everything in the centre is named after the female sex. There are Emily Gap, Jessie Gap, Trephina Gorge, Glen Helen, Ruby Gorge and Claraville. I wanted to meet Molly from Old Andado station, assuming she'd be an older, wiser version of Mrs Smith and Mrs Eyres. She was staying in her small bungalow in Alice, so I looked her up in the phone book. When I rang, she grudgingly invited me over, but it was a mistake. She was crabby and tired of her role as a woman of the outback. 'We were two young people kicking off in life, and we were given the chance to run a station. Who wouldn't have jumped at it?' she said, already in a well-worn groove. 'We worked like mad things to make it a success. Then my husband was killed. Well, that was that.' I nodded, thirty-year-old condolences seeming tame. 'The pioneering books never talk about the women who came up here, do they? But I wasn't the first. Australian men played their games. The women who came here made the real sacrifices, scavenging around, trying to make a home, even droving the sheep. The woman's tale has got lost. Yet she was the hub of the wheel, expected to stand her ground while the men turned. No wonder so many women went mad . . . the loss of companionship, the dwindling of the marriage. . . The men gave their all to the bush.'

So was she happy before he died?

'My husband told me my job was to bring up the children and educate them. I saw Alice Springs once a year if I was lucky. But I learnt to ride. Of course my life was unhappy at times. But the women shopping in Woolworth's still look miserable to me.'

There were family rumours that Todd had once bought two properties. Alice, it seems, was lucky she wasn't dragged there to cook boiled beef. But at least in this wilderness she might occasionally have had her husband's undivided attention. If Todd was in Adelaide, he'd spend twelve hours a day at his beloved post office, striding up and down the balcony, watching the letters being sorted. When the mail ships came in from Europe, he would sometimes spend twenty-two hours at a stretch passing on messages to Sydney alongside the other telegraphists.

The weather reports interrupted breakfast, his evenings were spent addressing meetings of the Adelaide Philosophical Society, and at night he reverted to being an astronomer. On Sundays he was a pillar of the church in the morning, and read research sent to him from Greenwich Observatory in the afternoon. Weeks were spent inspecting lines, an excuse to ride through the countryside with a swag and a whisky flask. Days were consumed visiting rural post offices. Alice was left at home to look after the poultry yard, dairy and vegetable garden, and supervise the children's education, meals and washing. One of them was almost always ill with measles or mumps and needed nursing. Major-General Symes, in his notes on Todd, only mentions Alice as 'a competent housewife'. But this was no mean feat when the observatory was often packed with visiting academics, her own six children, three of her brother-in-law's children and her mother, while she was trying to keep up appearances on a civil servant's salary. No wonder she used to retire to her bedroom with a headache.

I went to the Old Timers' Home, which, like almost every old people's home, was filled with octogenarian women fussing over a lone septuagenarian gentleman. I wanted to know if anyone had any memories of my family.

I joined Peg and Pearl. One remembered Lorna having auburn hair and an emerald-green silk suit. 'She came to the outback as the guest of honour at some party, I think,' said Peg. Back in Adelaide I checked through the library. The *Centralian Advocate* had recorded the event. 'Miss Todd stated that it was the greatest thrill of her life to take the flight from Adelaide, and that it had always been her longing to come to Alice Springs. Although Miss Todd has been seven times to England, this was her first flight,' the paper reported. The front-page story continued: 'Miss Todd has two nieces named Alice at present in England whom she dearly hopes will come to Alice Springs when they visit Australia later this year.'

Every generation of Alices had visited the town except for the original Alice Todd. I had asked Pat if she could remember the eighty-two-year-old Lorna going in a plane called the 'Sir Charles Todd'. 'Of course. There was a terrible fuss. She hated leaving Adelaide, the flight was the most frustrating event in her life. She couldn't walk around like a boat, so there was no socialising. She said she'd looked down at all that red earth and thought: "Why on earth did my father bother putting a wire over that?" Pat had also been to Alice. 'For the centenary. I went to judge the carnival floats, most of which were Mickey Mouse or Snow White, but one had the Todds. He was perfect, right down to his beard and white breeches. Alice was far too blonde and brazen. But I still gave them first prize. I don't remember much more except that I stayed with the bank manager and laddered my stockings.'

Few other relatives had made the pilgrimage, although a group of grandchildren did gather in London in the early 1970s. They were brought together for an exhibition at Olympia called 'A Town Like Alice'. The *Centralian Advocate* described it as 'an Aladdin's cave of canned fruit and other delicacies which bears about as much relation to Alice Springs as Paris does to Timbuctoo'. It added,

> Of some eight surviving grandchildren, six were present, the odd thing being that all but one of the second generation have returned to the country which their grandparents left a century ago. Of Alice's children, only one daughter, Lorna, is alive. She lives in Adelaide. One of the grandsons, Sir Lawrence Bragg, left at 16 and still remembers his grandmother's apple fritters.

My mother still has the recipe, written in Alice's hand.

3 apples
Frying batter
Hot fat
Sugar
Lemon

Peel and slice the apples into rounds, take out the core with a small round cutter, make frying batter, and flavour with lemon juice. Dip in the pieces of apple, plunge into plenty of hot fat, and fry till a good colour. Drain on kitchen paper, pile high on a dish, and sprinkle well with sugar; serve very hot with cream.

If Todd and Alice had moved to the outback, would their grandchildren have felt more Australian? Would they have been running properties instead of Cambridge laboratories? Todd thought of himself as Australian, Alice didn't. She never saw the new country, just a pastiche of the old. I wished that Todd had brought her up the line once in all his travels so she could have witnessed the true colours of the land to which she pledged herself at eighteen.

While Todd was running up and down to the Peake in the first six months of construction of the overland line, Alice was trying to keep up morale at home. The colony had borrowed £170,000 to fund the enterprise and, within days of the men leaving, the town was jittery. More and more politicians warned that the Great Project was doomed. There was panic that there would be a run on the bank. Alice had to keep calm, although the family finances were also rocky. They had had to sell her pony cart and let the second maid go. That first year, Alice valiantly made the tea-party rounds to ensure that no one was whispering about her husband; she relayed entertaining anecdotes and refrained from mentioning any outbreaks of scurvy. She wrote notes to the wives of the teamster men, urging them to be brave. After a time, the town almost settled back into a routine.

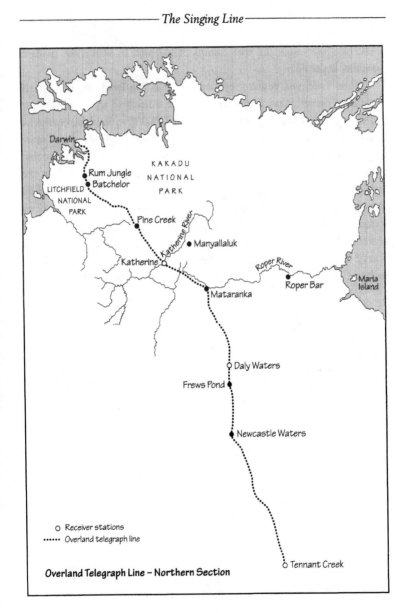

Darwin

KAKADU
NATIONAL
PARK

Rum Jungle
Batchelor

LITCHFIELD
NATIONAL
PARK

Pine Creek

Katherine River

Manyallaluk

Katherine

Roper River

Roper Bar

Maria
Island

Mataranka

Daly Waters

Frews Pond

Newcastle Waters

o Receiver stations
••••• Overland telegraph line

Overland Telegraph Line – Northern Section

Tennant Creek

15

· — — — —   · · · · ·

Just Wait for the Wet

Todd had been so engrossed with the bottom half of the line that he hadn't thought about the Top End. He'd watched the ship sail off to Darwin six months before with few concerns. No discouraging missives had yet been sent back. The gossip on a returning barque was that ninety miles of poles had been erected in the first seven weeks. The two successful contractors, Joseph Darwent and William Dalwood, had been given four weeks to get ready to sail to Darwin. No one in Adelaide understood the tropics and so they hadn't been warned that their heavy draught-horses would have problems in the damp. All they knew was that a few of the rivers flooded in the summer and further south the country was drought-ridden. Todd had instructed the government overseer, another McMinn, William, a twenty-six-year-old architect, to ensure that camps were made on high ground so that the supplies wouldn't become swamped.

Four hundred men applied to go to the Northern Territory, eighty were chosen. They were to be paid 4s. 6d. a day and rations, with a £1,000 shared bonus from the contractors if the line was completed on time. Darwent and Dalwood put a man called W. A. Paqualin in command of the party and chartered the *Omeo* to sail the men round the coast. Seventy-eight horses and ten bullocks occupied much of the space below deck. The upper decks were filled with hay, wagons and provisions, leaving little room for the eighty men, six officers and the government overseer and his team. The *St Magnus* barque was sent as a backup, packed with wire and insulators.

Dalwood had insisted on taking his wife and young son to see the work inaugurated. One thousand spectators turned out to listen to Todd's speech and give three cheers for the expedition as the boat cleared the wharf on 20 August 1870. Ralph Milner, an enterprising drover, set out the same day to try to take a mob of sheep from Adelaide to meet them at the top. The next Sunday, a service was held in St Paul's Church to pray for divine protection for the men of the northern line. Their prayers proved to be in vain.

The first disaster was only just averted by the *Omeo*'s captain. When he picked up more coal at Newcastle, the ship was dangerously overloaded for its 5,000-mile trip. The wind started to get up as they reached the Raine Island Passage through the Great Barrier Reef. The ship, a large vessel with both steam-propulsion and sails, grated against the coral and was only just floated off when the tides changed. The horses started getting fevers and ten men went down with malaria. The captain stopped at Booby Island to collect the post left by other vessels, and sailed into Port Darwin after twenty-three days.

The fifty inhabitants of this tiny town were stunned. No one had told them about the Grand Project. There hadn't been time. They had assumed the line would go through Queensland and that their isolated settlement would eventually wither. The Resident, Captain Bloomfield Douglas, had only arrived three months earlier. He was about to quit when Paqualin sought him out in his office tent. Now he wanted to see the project through. Darwin, he realised, could become the ear of Australia. Since there was no proper jetty, Paqualin was told the men would have to swim the horses ashore and float the bullock wagons and provisions to the beach. The locals, a mixture of whites, Chinese and Indians, were surprised at his flimsy equipment. 'Just wait for the Wet,' they warned.

Six days after the *Omeo*'s arrival in the Northern Territory, the entire town gathered at four p.m. to watch the first pole being planted on the corner of the newly created Cavenagh Street and the Esplanade and the Resident announced a public holiday. The *Omeo* was draped in bunting, the Aborigines were given coloured skirts, and the flag flew over the residence tent.

Miss Harriet Douglas, the Resident's prettiest daughter, performed the opening ceremony in a white dress. She took the new rammer and rammed the earth around the pole 'with most artistic skill', according to reports, saying, 'In the name of Her Most Gracious Majesty Queen Victoria, I declare this pole well and truly fixed.' Mrs Dalwood, the contractor's wife, broke a bottle of wine on the pole and they sang the national anthem. The gunners on the fort fired a twenty-one-gun salute that lasted an hour. Three cheers were given for the Queen, Miss Douglas and the Governor and three more for the men. The Aboriginal tribe called for drink to celebrate but were refused. One officer commented: 'The natives will soon be civilised when they think nothing can be done well without having a wet over it.' The *Omeo*'s captain held a small evening reception, where Miss Douglas and her sisters danced with the officers and drank South Australian champagne. A few days later the *Omeo* took the Dalwoods back to Adelaide.

The problems soon began. It was 250 miles from Port Darwin to the Roper River, where another camp was meant to be based, and no one had any idea of the terrain. The second party could not set out until it knew where it was heading. So it was decided that groups A and B would both work on the first section until the land had been surveyed. In fifty-four days the men built eighty-two miles of line. The weather was humid, they were soon covered in bites and sores, and they begged to be allowed to discard their shirts in the mangrove swamps. S. W. Herbert, at eighteen the youngest member of the party, wrote in his diary:

For provisions we had tinned meat, preserved potatoes, dried in rough particles like tapioca, and dried vegetables in pressed cakes, flour and biscuits. The cooks made 'Hash me grandre' stew, by putting dried potatoes into a half-bucket of water, adding 8lb of tinned meat, and stirring well. This stew when put into tins and baked, with preserved vegetables, makes Lobscouse . . . I had a lovely pair of boils on the back of my neck, with prickly heat as well. Being tormented by mosquitoes made life almost unbearable. The climate was hot and muggy, and the flies more troublesome than expected. Blowflies were a perfect nuisance. Axemen and others complained that their moist woollen shirts became flyblown, and live objects would soon appear.

Then it rained. The men all swigged a beer to celebrate. But the showers didn't stop. Their tracks soon became quagmires; their attempts to stick to higher ground made no difference. The backup teams were cut off from the forward parties, the sugar and salt got weevils, the flour was sodden and the advance explorers came back with depressing stories of the interior. Led by the government surveyor, G. G. McLachlan, they had gone round to survey the Roper River and try to take stores up that way. The river, they discovered, was just navigable, but the countryside was barren. The second mate, Jack Reed, had taken a party to find fresh water in his small rowing boat. One night they heard a piercing scream and saw Reed disappearing over the side. A crocodile had grabbed his leg. They never recovered the body.

The men on the line were getting twitchy. No one had mentioned crocodiles. One horse was killed from a snake bite and McMinn found a particularly dangerous viper in his boot which, fortunately, had been so squashed by his stomping that it had been unable to attack him. The Aborigines kept coming into the camp, and one man was nearly speared when he led his horse to a nearby waterhole. The Aborigines also tried to rip down the new line. The whites thought they were the first Europeans on the land until they discovered some old dray tracks and concluded there must be ghosts. An escaped convict was the more likely explanation.

New provisions arrived from Adelaide in the *Tararua*, but they were bogged down just outside Darwin. The men were soon stranded in the interior by the rising rivers. They soldiered on, eating their tinned bully beef. By Christmas Day 1870, as Todd was carving the turkey with his family in Adelaide, his men had reached Pine Creek and had erected 180 miles of line. They slaughtered a bullock and held a sports day. Herbert wrote:

> We had foot races, jumping, putting the stone, grinning [sic] through a horse-collar and bobbing at the duff. The latter took place on the bank of a lilly pond. The duff was covered with a sticky sauce and suspended by a string from a branch. Competitors were invited to take a bite at the pudding, at the risk of falling into the pond. Prizes were given, and being a teetotaller, I was surprised to get a bottle of Jamaican rum. This was a joke, as Tommy, our champion step dancer and no teetotaller, was given a bottle of lime juice. Tommy interviewed me about an exchange, but, as I held the trump card, he had to ante up some tobacco as well.

Then it really rained, and the men discovered the meaning of the Wet. The season lasted five months and the start was signalled by monsoon clouds sweeping in from the ocean. At Port Darwin, the first settlers had got used to an average rainfall of sixty inches. Pine Creek gets about fifty inches of rain and the Katherine forty inches – still enough to flood whole valleys.

The land was soon one huge lake. No sooner had they dug a hole for the poles than it filled with water. Axes rusted overnight and the wire coils solidified. Herbert's mood quickly changed. 'The flour was dirty and full of weevils. The bread looked like currant cake when it was baked. The weevils swelled and burst in the dough in the camp-oven heat, and discoloured the bread . . . No one would touch the rice, as it was alive with tiny grubs. Until now, the men had laughed at the idea of the Wet Season stopping work, but now they knew better.'

The supply teams had an easier time. Having got stuck with the rum at the flooded East Finniss River, they went on a spree: the place is still named Rum Jungle.

Paqualin insisted that the men forge on. When they reached the Katherine River at the end of January 1871, he made them build rafts so they could ferry the equipment across. The operation took the whole of February. For a week the men planted fewer than ten poles a day. Then they mutinied. It hadn't taken them long to realise that no supplies were following them, and they had run out of tea, sugar and alcohol. A deputation went to Paqualin after fifty-six men voted to strike. He briskly told them that any man who quit would lose his pay and any share of the bonus. Eleven men insisted on being given a cart and horses to attempt to return to Darwin. They slaughtered a cow and had a barbecue along the way, convinced that they had made the right decision. The others stuck it out.

Meanwhile, the government overseer, McMinn, was nowhere in sight. He had been stranded in Darwin and lost a horse trying to ford the rivers. When he eventually reached the advance team, he was shocked. He wrote a formal letter to the contractors, saying that he found their progress unsatisfactory. Paqualin was angry, defending himself by pointing out that he had poled over two hundred miles of line and wired a hundred miles of pole. McMinn worried about morale among the men. But he had to press on to meet Ross and Giles coming up from the centre, so the argument would have to wait. He searched for the central party for a fortnight, having no idea that Ross hadn't yet found a route through the MacDonnells. Eventually he buried some stores for the men under a tree in some glass jars and returned to Paqualin's camp.

The camp was deserted when McMinn rode in. The remaining men had all mutinied and insisted on returning to their depot on the Katherine River. Many had already set off for Port Darwin. When McMinn finally traced Paqualin, he was livid. A small, excitable man, he immediately insisted that the whole operation was off and declared the contract void. According to the small print written by Todd, as government overseer, McMinn was allowed to take possession of all the materials and carry on the work himself if he decided that not enough progress was being made. McMinn also had points to score, having tendered for the line himself and lost the contract to

Darwent and Dalwood. Before the expedition set sail, the two men had pointed out in Adelaide that this gave McMinn far too much power. But Todd thought the strict wording of the agreement, stating that McMinn could act only in an emergency, would prevent dirty tricks.

Todd was now sent a letter by McMinn saying he was taking control. 'From the experiences I have gained of the country traversed and of the nature of the work, during my supervision of the contract, I am of the opinion that it is possible to complete the line within a few weeks of the stipulated period,' he wrote. But, as Paqualin argued, why would McMinn succeed where he hadn't? McMinn was adamant that the men only needed better leadership. Paqualin pointed out they would lose the generous incentive bonuses offered by Darwent and Dalwood. McMinn said he knew that fresh supplies were soon arriving on the *Marie Elizabeth* supply ship, but Paqualin said they were for his operation.

The two men squabbled all the way back to Darwin. Invited to dine at the Residency, 'they couldn't speak a civil word'. When the *Marie Elizabeth* sailed into the harbour, they both rushed to the captain to commandeer his stock. But out of the twenty horses only thirteen were still alive and the amount of wire and insulators was insignificant. Now the tables were turned on overseer McMinn and he no longer had the upper hand. He had forcibly taken over responsibility for the operation, but it was clear there was no way he could fulfil the brief. If Paqualin was returning to Adelaide to argue his case, McMinn decided he must go too. The *Gulnare* was leaving in a week. They both scrambled to get berths on it, leaving a mess behind.

Seeing their bosses disappearing, fifty-five workmen quickly booked their tickets home. Only twenty-seven still remained when the boat set sail on 6 June 1871. In seven months the line had to be open or South Australia risked bankruptcy. The trip can't have been a social one, with Paqualin insisting his bosses would sue the government. When the men reached Adelaide a month late on 5 July 1871 there was uproar. There were only six months to go before the deadline expired. Superintendent Todd was at Strangways Springs, near Lake Eyre, when the news was flashed to him on the wire which had just been fixed up. He rode flat out to Port Augusta, a ten-day journey of over 330 miles, and took the first boat to Adelaide to join the frantic consultations that were taking place in Parliament. But he was too late. The government had already taken its decision. Assistant railway engineer Robert C. Patterson had been selected to take control of the north. He was only twenty-seven-years-old and the parliamentary records describe him as 'the well-known gentleman'.

16

·———— —····

Water on the Line

Back in Alice, we decided to follow Paqualin's line in the rainy season to judge for ourselves why the expedition had foundered. This, we thought, would give us some sense of the conditions, but we would have the advantage of windscreen wipers. Everyone had told us that the second half of the trip would be easier. The vegetation was lush, the waterfalls were spectacular and the gorges had some of the best white-water rafting in the world. The Top End was meant to be stunning in the wet season, with Jesus birds walking on the water and fields of water-lilies.

Much of the route would be straight up the Stuart Highway. It sounded quite easy. How could we get lost or bogged on a tarmac road? Any detours would be well signposted. Ian helped with the re-equipment exercise. Knowing the ground would be permanently soaked, we decided to park our swags and Toyota with him, and get a four-wheel-drive camper-van. This would allow us to sleep above the puddles in a squashed-up compartment in the roof. The van had a mini-fridge, a cooker, two chairs and a table. There was a light for night-time backgammon and a hook for our Akubras. The camper-van still didn't conform to Ed's ideal of driving across the plains in his open-topped Cadillac, but I was quite at home in my snail's shell, having spent childhood summers crossing Europe in a Volkswagen bus. We could take chilled Chardonnay and steaks and pick up fresh milk for our cereal. It would be luxurious compared with the first half.

With only two weeks left, we planned the route meticulously, working out exactly where the advance party would have landed on the Roper River, and where the first man died. We wanted to charter a boat to navigate the Roper. The hard part was jettisoning our well-loved junk. We binned squashed petals from desert roses, cockatoo feathers, chewed-up tapes, two punctured tyres, our oily rope and the untouched emergency Pot Noodles. For extra precautions, we refilled our forty gallons of water in case it stopped raining. Ian sent us off with a farewell party. His dogs were dressed up in sunglasses and headscarves and we drank daiquiris until dawn.

Next day, lurching from side to side in our new high-rise, we wistfully headed north past the Chicken McNuggets. It was already late afternoon and we were nursing hangovers. We stopped at the tourist office to tell them we were on our way. The woman at the desk looked concerned. 'Hold on a sec.'

A man in shorts and long white socks came out. 'I'm afraid you can't go.'

'Of course we can. We know all about the snakes and the road-trains. We've done Adelaide to Alice across country,' I said confidently.

He shrugged, 'Well, try if you want, but the road's closed.'

'We're going along the Stuart Highway, not overland,' I explained very slowly.

'Well, the Stuart Highway is closed,' he replied even more slowly. 'Due to flooding. It's the worst for ten years. You won't get beyond Katherine. Forget it. The gorges are overflowing, the whole of the national park is a lake, there are plagues of frogs and you'll only see the waterfalls by helicopter.'

I didn't care about waterfalls. What about the Roper area?

'You'd need a high-powered speed boat to get anywhere near the settlements and those will be pretty booked up right now by refugees bailing out.'

Two hours later we were back at Ian's house. 'I know it's all flooded,' he commiserated. 'The tourist office rang after you left and I just heard it on the radio. Dramatic, isn't it?'

'Not really,' I replied, unpacking our steaks. 'How long do you think it will take to clear?'

Ian didn't pause. 'Oh, it'll settle in now, it will probably pour for weeks. Perfect for Ed's bronchitis.' He winked at Ed. It was a mild thirty-two degrees outside, the sky was cloudless and the sun shone. We hadn't felt a drop of rain, yet three hundred miles up the track it was pouring down. I couldn't believe it.

Ed was more pragmatic. 'Why don't we call it a day and come back in August when it's dry again? This isn't the last holiday of your life and I can wait for another round of toasted sandwiches.'

I couldn't bear leaving something half finished, but my co-pilot was stubborn. 'Do you really want to spend two weeks watching the rain coming down on the bonnet?' he asked.

'At least it's authentic,' I said.

'We won't have a hope in hell of finding any traces of the original expedition,' Ed added.

Ian backed him up. 'It'll give you another excuse to come back to Australia and have a beer on us.'

So we returned our pristine camper-van and picked up the Toyota from Ian. The idea of driving back to Adelaide along the Stuart Highway was too depressing, so we decided to fly straight home to Britain and send the car back south by train.

Our final trip was to the railway station to send the Toyota back to Adelaide. The Ghan train, which still connects Alice to Adelaide, was fourteen hours late due to the heat and had only just pulled in when we arrived. Children were sitting on their trunks waiting for a new term at boarding school, and Mr Rogers, the self-effacing stationmaster, was pacifying a honeymoon couple who wanted to get home. He had come to Alice twenty years ago as under-stationmaster when three trains went through a day. Now only backpackers, railway fanatics, romantics and children use the twice-weekly service. Articulated trucks drive the orange juice and cereal for a late-twentieth-century breakfast up from the coast.

Mr Rogers adored the Ghan, named after Todd's Afghan camel-men. He had first travelled on it as a child and no other train journey compared to watching the steam curl in front of him across the outback. He let us take a quick look at the train while it was in the sidings. The smoking rooms, dining cars and the honeymoon suite still had the original interiors. Alice, he thought, had changed far more. When he first arrived, people would come out on to their porches to watch a plane fly over. Now there were flights from Adelaide to Darwin every few hours.

As Ian drove us out through Heavitree Gap to the airport, I felt guilty that we'd abandoned Todd's expedition as it was reaching crisis point. Todd must have been frantic when he returned to Adelaide and was told that the northern operation was turning in to a catastrophe and, even worse, it was out of his hands. The government was taking direct control and Robert Patterson was now commanding the emergency expedition.

Our prospects looked just as bleak. Both of us were moving to new jobs and had no idea when we would make it back. But as we touched down at Darwin, we realised we'd made the right decision. The rain was coursing down the runway and it took half an hour before the plane could land. Skidding down the stairs, we were soon drenched with warm rain. As we sat shivering and waiting for the plane back to England, I knew we would have gone mad driving through weeks of wet. The mould was already forming on our rucksacks, the ants would have got into our breakfast cereal, and Ed would have caught pneumonia.

17

·— — — — — —··

The Defeat of Fame

For the next six months I was stranded in England starting my new job, appropriately enough at the *Daily Telegraph* newspaper. Sitting in the offices at Canary Wharf in London, I was only ten minutes' drive from where Todd was born, but half a world away from Alice. I occasionally read the news reports from Australia about killer sharks, children who'd swallowed pairs of scissors and survived, and the growing ascendancy of a woman called Pauline Hanson. The owner of a fish and chip shop, she had become the darling of the Queensland rednecks for speaking out against subsidies for Aborigines.

It was easy to shake off the red dirt when surrounded by concrete and tarmac, but at the back of my mind I knew that I'd left the Todds in the lurch. Whenever I had time, I would go to Australia House on the Strand and sit in the library leafing through photostats of diaries, reading Adelaide's the *Advertiser* and looking through my notes on parliamentary records. Surrounded by Australian accents, I would slowly link up the pieces, before wandering down the street to meet Ed at the American Bar of the Savoy Hotel. It was a far more civilised way to travel across the outback. But I missed skidding our four-wheel-drive over the dunes, the triumph of finding a can of peaches that hadn't passed its sell-by date, and the satisfaction of discovering an original wooden pole.

Thirteen months had passed since Todd told the men to start digging. By the beginning of July, he only had six months left in which to salvage his reputation before the deadline ran out. Todd privately told his old friend McGowan, who was on a visit from Melbourne, that he was convinced that the whole project would collapse. They had to find a way of putting up the poles in the Wet or he would get the blame. He insisted he would be thrilled if the assistant railways engineer Robert Patterson managed to turn the fiasco around, but McGowan must have known that if Patterson succeeded where Todd had failed his friend's career would be on the line.

Todd had never looked in worse shape. He was barely bothering to eat, his mouth ulcers had flared and he had dislocated a shoulder helping to ram a pole into the ground. Invitations to dinners and dances had dried up and he was no longer called upon to give the sermon in church. He'd spend hours shut away at his desk in the post office, often sleeping there overnight, and even his staff began to avoid eye contact. If Todd lost his job in South Australia, there would be nowhere to go. Who would employ a telegraphist who had not only failed to link Australia to the motherland, but had also gambled away his colony's treasury? Certainly no other colony would take him on, and his reputation in England would be as muddy as the land in which his hopes had sunk. Alice, who was floundering under the snubs and gossip, urged Todd to pray. But he brushed her off, determined to wrest control from the politicians and have one last stab at making his project work.

Todd started with the Governor over tea, and worked his way down through the politicians to the postal workers, trying to convince them that the project was not going to fail. He pointed out that the line to Alice was nearly completed, and tried to convey how proud he had been as he rode down the poles, stepladdering his way back to Adelaide.

The teams had settled down to a routine. The southern section would reach the Peake by October. The central section was also well advanced. The camels were already being sent back and the operators were making their way up the line to test the wires and establish the repeater stations. Alice had already become a small town. By the end of September 1871, eight hundred miles of poles would be ready and the path through the MacDonnells would be well established. The batteries had been installed, and men were already amusing themselves sending messages back to their families.

The route was now well worn from the carts, though venturing off it was still hazardous. When Knuckey, the overseer of Team A, ran out of salt at his camp fifteen miles from Charlotte Waters, he went cross-country to Dalhousie Springs with a man called Fitch and two packhorses. On their way back, Fitch got rheumatic fever. Knuckey tied him on to his horse and led him along, but progress was slow, and the Aborigines were circling them. They slept out that night and the next morning Knuckey set out alone, leaving Fitch with a revolver under the tree. When the rescue party arrived twelve hours later, Fitch was nearly dead. Flies were feasting on his oozing eyelids and he'd tried to shoot himself as the crows pecked at his sores. But he recovered. A man called Palmer was not so lucky. He died of consumption a week later near Central Mount Stuart. Another man blew his hand off and the stump went gangrenous. Todd stressed that they had lost only three men while stringing his line to Australia's navel, but, as the politicians pointed out, there was no point in constructing a telegraph system that ran out in the central scrublands.

Todd tried to explain why Patterson's approach would fail. Not only were his supplies still inadequate, but Patterson was a railway engineer and had no clue how to marshal men, no previous experience of the tropics and no understanding of telegraphs. To make matters worse, Todd now calculated that this heavy 'rainy' season in the Northern Territory was not atypical.

The government didn't want to listen. It was going to give Patterson every assistance to complete the line. Money was no object. Patterson was offered a lump sum of £300 in addition to his regular salary, and a bonus of £1,500 if the line was completed by 31 December 1871. If he managed it by 15 January he would get £1,000, and even if the whole project was delayed to March he would get £300. Adelaide was surprised by the size of the reward. Todd must have felt bitter that he had been given no such incentive.

Patterson seized the moment. He advertised for recruits in the *Advertiser* and by the next morning 525 new volunteers were queuing up outside his office, curious to beat a path into the interior of their new nation. Patterson selected eighty-seven men.

The new boys would set sail at the end of July with six vessels – the steamer *Omeo*, which they picked up in Melbourne, and the barques *Bengal*, *Golden Fleece*, *Laju*, *Antipodes* and *Gulnare*. McMinn had reported that draught-horses were unsuitable, so Patterson was given 170 light horses and 500 working bullocks – Premier Hart had gone personally to Melbourne to buy the extra animals.

The Adelaide shopkeepers were making a handsome profit, so they were happy. The saddlers made 153 new sets of wagon and express harness, 600 pairs of hobbles, 1,500 straps and 13 saddles. It was obvious that the state had now gone way over its original budget of £120,000, but no one was counting anymore. All that mattered to the government was that South Australia should not be humbled. The *Advertiser* was a lone cynical voice, and one of Todd's few remaining champions. The editor wrote: 'We are not sure . . . if it is not somewhat infra dig for the Premier to be personally superintending the buying and selling of government stores and selecting horse and cattle . . . Mr Todd, we need hardly say, is working night and day with his usual energy.'

Todd was allowed back into the consultation process. His recommendation was that the provisions for the new offensive be divided in two, with Patterson going to Darwin and his deputy, Walter Rutt, going up the Roper River and bringing much needed supplies straight to the interior. At first Todd thought that they had accepted his proposal, but after the ships set sail the government cancelled the plan, telegraphing ahead to Sydney with the message to insist that all men went to Darwin. They told Todd that they had decided it would be more sensible to keep the teams intact. But Todd knew the real reason his proposal had been rejected. The politicians were worried that a port on the Roper might become a rival to Darwin, which they wanted to establish as the pre-eminent town in the north. Now politics was conspiring with the weather to wreck the project.

There was nothing much Todd could do back in Adelaide except urge the men of the middle section to finish their work as soon as possible and head north to help. To encourage them to keep digging, he sent quantities of sugar, tea and rum up the line. Todd's temperance tendencies were proving no match for his ambition.

But his real problem was Patterson. He'd never liked the man, and took care to avoid him at social occasions. Patterson did everything by the book. Todd, although himself a perfectionist, thought him pedantic. They were both concerned about their reputations as public servants, but where Todd jollied the men along, Patterson was cynical and gloomy to his juniors and unctuous to his seniors. He was also obsessive about both his wife and his religion. His clammy-handed outpourings about his wife, Elsey, and his endless genuflecting unnerved Todd, who never mentioned Alice to the men. Patterson was always licking his thick, red lips and toying with his beard, he had a tic in one eye and towered over Todd, to the super-intendent's irritation. But worst of all in Todd's opinion, he was a natural pessimist who was constantly warning South Australia against being 'too big for its own boots'.

While Patterson gathered his supplies, Todd insisted that the politicians make it clear that he had not been sacked. A notice appeared in the *Register* newspaper on 17 July stating that Todd was in charge overall, but that Patterson would be 'unfettered' in his work in the north. The Adelaidians weren't happy. The colony was spending more and more for what looked likely to be a dismal flop. One newspaper wrote: 'The whole concern looks like a picnic on a large scale.' Others penned songs and ditties, a favourite being: 'The Northern Line I guess, will end in an alarming mess. And if I rightly this infer, We'll have to thank the Treasurer.'

The men sailed from Port Adelaide on 27 July 1871. Todd sent them off with a rallying speech, saying: 'Credit is in your hands . . . you are in fact so provided that if you do not do the work you will come back as a lot of disgraced men . . .But I am certain you are prepared to do your work and maintain the credit of South Australia.' The implication was that Patterson should find it easy to complete the project, but Todd knew that his rival was doomed.

Patterson arrived in Port Darwin in August facing exactly the same problems as Paqualin a year before. He had only three months to get everything up and running before the wet season started. The new stores didn't help. Over a hundred of the five hundred bullocks had already died on the voyage. Some were so parched on arrival that they drank too much seawater when they swam to shore and died. Others soon fell ill on the meagre tropical grass available at the end of the dry season. Many horses disappeared into the bush in search of feed and never returned.

In the four months since McMinn and Paqualin had fled, it looked as though no poles had been erected. The overseer left in charge during the dry season had only managed to rethread sixty-nine miles of wire and excused himself by saying he wasn't sure which route he was supposed to proceed along. Worse, many of the men had gone off gold-prospecting instead of building the necessary bridges for the Wet, and refused to be recalled.

To the men's surprise, Patterson's first preoccupation was to name the newly discovered tributary of the Elsey River Birdum Creek, 'Birdum' being his pet name for his wife. He then turned his attention to the route, and decided to adopt the shorter path, proposed by Giles and Ross, who had finally made their way up after their explorations. He sent the new men up the line as quickly as the gear could be assembled. John Little, the new Darwin telegraph master appointed by Todd, arrived two weeks later and tried to jolly things along. Todd called Little a man of 'great integrity, intelligent determination and a good administrator'. He was also a telegraph engineer, unlike Patterson, and answerable only to the Postmaster-General.

Patterson and Little soon clashed over the proposed route, with Little adamant that the men should concentrate on the Darwin end of the line, rather than risk getting bogged down in the interior. Little wrote in his diary, 'Terribly anxious about the line and feel certain that Patterson's plan will be a failure.' Patterson, normally the most cautious of men, felt pressurised by the lure of the bonus, but the men coerced him into a compromise. The push into the interior would continue, though at a slower pace. Whatever Patterson did, he felt jinxed. 'My prospects are very gloomy,' he wrote after a fortnight. Another entry began, 'It is a matter of great regret not to be able to report favourably on my prospects.' Little and the men had taken an instant dislike to Patterson. Little wrote back to Superintendent Todd: 'Mr Patterson has done everything in his power but a strange fatality attends all his arrangements.'

Most of the arduous work would be south of the Roper River, and Patterson soon realised that Todd had been right. Taking a boat up the Roper would be the only way to get supplies to the men when the rain began. So the *Gulnare* was dispatched up the Roper, packed with provisions. Just as Patterson was beginning to relax, its cutter returned with the news that the schooner had grounded on a reef in Clarence Strait, only thirty miles away, and would not budge. Patterson had no alternative but to commandeer the only remaining boat, the *Bengal*, to take off the stores and proceed to the Roper.

As the dry season ended, the men were making slow progress down towards Alice, trying to dig wells along the way to ease their thirst. The overseers sent messages back saying that the animals were dying in their dozens and the new equipment wasn't up to scratch. By the end of October, when the Dutch warship *Curacao* dropped in on Port Darwin to snoop on the progress of the wire, Patterson hadn't erected a single pole and there were only two months left before New Year's Day 1872.

Patterson asked the *Curacao* to take dispatches back to Todd. His plea was for thirty more teams of bullocks or horses and more men. Todd finally got the message. 'The Telegraph Expedition is breaking down through excessive loss of stock by deaths at Port Darwin, through disasters attending the passage of the dry country and through knocking up of stock generally.' But the appeal had only reached Adelaide in December, too late to provide help before the rains. The men kept themselves going along the line singing: 'If the job isn't finished by '72, Patterson's bonus will look pretty blue.'

Then Patterson's confidence took a worse knock. The cable-laying ship, the *Hibernia*, had arrived in Port Darwin. In her hold was the coiled cable, ready to link Darwin with Java. The efficient Captain Halpin was in command of the expedition of electricians and engineers. As the three hundred new arrivals started to bring the cable ashore, they assumed the overland line was nearly finished, and they weren't just linking the world to this dead-end. They dug their trench along the beach up to the new cable house, and promised their cabling work to Banjoewanji, in Java, would be finished in less than a month.

The visitors asked about progress across the continent. Patterson was forced to lie, saying the poles had almost all been erected. Handing out the grog, he packed them off on the ship as quickly as possible. That night the cable was slipped from the 'blights' and sank into the ooze of the harbour to begin its journey across the Timor Sea. Patterson felt humiliated.

Little wrote a tactless missive to Todd: 'I am much struck with the magnificent and perfect manner in which everything on the *Hibernia* is organised and carried out. The size of staff, number of men, quantity of stores and material is immense. There is no waste or extravagance, but still a profusion of everything which renders success certain.' His views on Patterson by this time couldn't have been more different: 'Patterson attempted an impossible task . . . The attempt was made, the parties pushed out too rapidly and short-provisioned. They are isolated, with their stock in such dreadful condition, that not a pole has been erected. It is now a question of rescuing the men from starvation.'

Already suffering from heatstroke, Patterson headed for the bush to establish a base at the Roper before he had to answer too many more embarrassing questions. But this was not before he had insisted on posting the protesting Little on a dangerous voyage back to Queensland, just as the monsoon season was starting, so he would stop meddling. Patterson then set off for the camp on the Katherine River. It took him three weeks to reach, and he discovered that Little was right: not one pole had been erected since McMinn had annulled the contract. The dry season had passed without progress and there was still a gap of four hundred miles. Patterson momentarily revived when he heard about a flock of sheep being driven across Australia by the Milner brothers. One brother had been killed by Aborigines, 3,000 sheep had been lost to poisonous plants, but there was still a remnant left which had crossed the entire continent and was approaching Katherine and the starving men. At least there would be fresh meat.

By 3 December his mood had flip-flopped again and he considered resigning. A week later he wrote: 'Prickly heat nearly drives me frantic. Flies and mosquitoes are unbearable. I am unutterably weary of the whole thing. Can see nothing but blackness and suffering ahead. Fear Expedition must collapse.' On 25 November, with one month to go before he forfeited his £1,500, he wrote: 'Rained four hours'. On 26 November, he added, 'Day by day I become more impressed by the hopelessness of my position.'

The *Hibernia* had done its job and local convicts had been used to prepare the new receiving huts in Banjoewanji. From eight p.m. on 19 November London could communicate with Port Darwin. This tiny population could 'talk' to politicians in Westminster in seven hours. But there was little point in being able to converse with a swamp. Britain had been promised a link to Adelaide, Melbourne and Sydney, not a muddy village. The British-Australian Telegraph company made it clear that Adelaide would soon be forced to pay the penalty – and not just financially. The company sent a letter threatening to extend the line round to Queensland. Even if South Australia's Great Project was completed, it was in danger of becoming an irrelevance, left to wither in the sands.

Patterson, travelling down to the Roper, had already got his wagon stuck in a muddy rut. The rains had begun again. Realising this meant defeat, he worked himself into a lather of self-pity, bitterness and despondency. One moment he would be shouting at his men, the next weeping in front of them. In long outpourings to Elsey, he blamed everyone but himself. Of the politicians in Adelaide, he wrote, 'I am going to christen my four wagonnette horses naming them after politicians, Hart, Milne, Blyth and Carr. And as I mostly drive myself, I shall have the satisfaction of yelling out their names all day long, touching them up with the whip when they prove refractory and thinking I am driving the Ministry over the most miserable country on the face of the earth.' His diaries reveal Patterson's self-obsession. 'It is raining piteously,' one entry says, 'and every drop falls on my head like lead. Plenty water now. Too much. The work is doomed. The lives of two hundred men are imperilled for want of supplies. I cannot be held responsible. I have done all that mortal man can do.' On 10 December 1871 he added: 'I have met with defeat of fame, but I hope in time to think of it with less pain than it gives me now.'

Not once does Patterson mention the conditions under which his men were labouring. But Giles, by then an experienced bushman, wrote in his diaries. 'The country was terribly boggy and we were obliged to lead our horses . . . even then some of them sank to their bellies in the bogs, with us up over our knees in slush and mud . . . the Elsey was more than half a mile wide . . . myriads of flies, sandflies and mosquitoes.' Giles was appalled by the lack of industry. 'Nothing was more depressing and perhaps more injurious for a climate like this than idleness and inactivity. It was the mother of discontent, disorder and dissentions, and contributed to disease in a climate where malarial fever was prevalent, and where numbers of men were camped together for a length of time without books or occupation of any kind.'

Patterson could think only about getting to the Roper, where at least he might find the equipment from the supply ship, the *Bengal*. He plodded on with a team of four and two wagons, covering only five miles a day. On

Boxing Day, they discovered the sheep, and Patterson immediately bought a thousand to be delivered to McLachlan's camp to avert famine. He also bought another more comfortable saddle for his increasingly bony frame. Christmas Day wasn't celebrated. 'Christmas dinner was bad damper and weak, mawkish tea,' Patterson wrote. The men abandoned their wagons and carried on with the horses. Every couple of days they would come across more telegraph men huddled together on the higher ground, with nothing to do except pick the weevils out of the bags of flour.

On reaching the Roper on New Year's Eve, Patterson discovered two teams. They were soaked through, and were living off fish. A jetty had been constructed but there was no *Bengal*. Instead, the teamsters had found a letter buried in a bottle beneath a tree marked 'Dig'. Only the *Bengal*'s small rowing boat had made it as far as the landing place and had left word that its parent ship was stuck at the mouth of the Roper and was having problems navigating the bars.

Patterson spent New Year's Eve squashing mosquitoes in his tent, reading *Pilgrim's Progress* and pondering his future. The line would have to be up and running the next day or, under the terms of the contract with BAT, Adelaide would start paying the price, and so would Patterson.

18

· — — — —   — — — · ·

The Superintendent Takes Charge

In South Australia, debate was furious as the deadline loomed. Hart's government had fallen the previous month, and the new incumbents had two choices: to scrap the whole project or to invest yet more money in it. It was now obvious that the delay would be considerable. Patterson was roundly blamed. One parliamentary report concluded: 'Our view is that he is far more likely to have exaggerated than to have understated his difficulties.'

Todd was now pushing the idea of a pony express. After all, the company had not stipulated how the messages had to be transmitted. The 394-mile gap between the two pieces of line at Tennant Creek and King River might now be small enough to allow ponies to travel with stops. But the government had other plans. It decided to send Todd north to the Roper River to make one last attempt, although they were not optimistic that he would fare any better than Patterson. 'He has no particular qualifications for transmission of stores in a tropical climate, but his general knowledge and authority as head of a department will be of undoubted service,' said the report.

Todd wasn't happy. His plan had been to let Patterson do the hard work on the Roper, while he took a ship up to Darwin and rode all the way down south, checking the line and boosting morale. Under the government's new scheme, he could well be away for up to a year sorting out Patterson's mess. If only they had sent Todd to the Roper six months before, he might have been able to make a difference. But the public servant knew he didn't have the luxury of choice.

The men further down the line were buoyed by the news that Todd was taking control again. Kraegen, a former stationmaster, decided to ride out from Charlotte Waters to meet his boss. He took two men, and set out overland. By the third day they had run out of liquid and were lost. Increasingly worried, Kraegen volunteered to go ahead and look for water. By the sixth day, the other two men had killed one of their horses and drunk its blood. Thus fortified, they had enough strength to find water. Kraegen was not so lucky. Parched and alone, he eventually found the poles about eighty miles from Alice Springs, wrapped himself around one and waited to die. A few days later, a linesman found his body. The new telegraph-master at Alice wrote in his diary: 'Christmas Day – Weather cool cloudy – Rain in evening and during night. The following is a copy of the certificate in reference to the death and burial of CWI Kraegen.' A century later, Kraegen's grandson found his makeshift grave with an inscription punctured in tin, 'perished for want of water', and erected a small headstone. Todd was distraught to lose his old colleague.

Two hundred miles away, however, the north was still a washout. Patterson, behaving increasingly erratically, decided to make a boat and float downstream looking for the *Bengal*. He had forty men on meagre rations and couldn't bear being cooped up in his tent any longer. So he took the wheels off a wagon and lashed a tarpaulin round the body to make it watertight. Then he added ten empty kegs to give the makeshift vessel extra buoyancy and made oars of bush timber. Only three men volunteered to accompany him. They had a hundred miles to cover, in the pouring rain, and if they didn't discover the barque they knew they would never make it back upstream but would instead slowly starve.

The makeshift boat was swiftly swept downstream, and by the next day they had sighted the masts of the *Bengal*. Triumphant, they boarded as night fell. Patterson was furious that the mild-mannered Captain Sweet hadn't made more progress up the river, and tossed aside all excuses that the headwinds were too strong. The next day he loaded two longboats with rations and sent them up to the camp. Four men in each boat were expected to row laboriously. Patterson remained on the barque to urge Captain Sweet on.

On 14 January, the *Bengal* was overtaken by a government cutter, the *Larrakeeyah*. Inside was a jubilant Little, bringing remarkable news which he knew would infuriate Patterson. The Postmaster-General, Charles Todd, had departed in person from Adelaide on 4 January, in the SS *Omeo*, bound for the Roper with supplies and provisions, eighty horses, and more workmen. A steam-tug, the *Young Australian*, would assist the *Omeo* up the Roper.

Patterson was livid. Returning with the *Larrakeeyah* to the landing stage and his base camp, he discovered that the river had overflowed its banks. No matter what supplies he brought up the river himself, it would be impossible to do anything before Todd arrived. Patterson was completely cut off from his construction crews in the interior. He would receive no credit if the line were ever finished, and Todd would become a hero if he managed to get the wire up and running. Patterson, who had risked his life on the Roper, would be relegated to a postscript.

Patterson's diary shows that he was almost deranged in his anger towards Todd, whom he accused of spreading malicious rumours. He scrawled: 'I find that Mr Todd considers my telegram from Port Darwin as unnecessarily alarming. He also attributes the partial failure to the assumed fact that the stock had been overdriven and not allowed the fortnightly rest after landing. I am extensively annoyed at the false assumption and the conclusion he has jumped to without a particle of evidence.' Patterson decided he had no option but to resign when Todd arrived.

*

With Charles tied up on the line, Alice was left to manage an increasingly complicated family life. She found the task a constant trial. Her closest confidant, Charles's niece Fanny, married Charles Davies in 1871. The match was perfect, Charles was the son of Dr Davies, the first medical practitioner in north Adelaide and one of the Todds' closest friends – but it meant that Alice was left to cope single-handed once again. Fanny's brothers, George Griffith and Charles Robert, may have idolised their uncle but were two more mouths to worry about. George begged to be allowed to help out Postmaster Little in Darwin, while Charles became a cadet in the Chief Secretary's office. On top of that, a nephew of Alice's, a William Squires, suddenly turned up on a boat one day. Charles arranged for him to train as a telegraph cadet.

That meant nine children and her mother, in an already overcrowded house. By 1871, when Todd set out on his trip to the Roper, Lizzie, his eldest, was fifteen. Close to her father, and already impatient with her mother's eccentricities, she was the practical one of the family and she studied assiduously at Miss Senna's School. Charles, fourteen, was a scholarly child, and Hedley, twelve, was the family sportsman, always begging to go with his father into the bush. They both attended St Peter's College, organised along the lines of an English public school, becoming boarders when Charles set out on his trip. Alice Maude Mary, now seven, was the tomboy and worshipped Hedley. Her dark hair was cut short with a straight fringe and she was always falling into the duck pond. The youngest, my great-grandmother Nina had inherited her grandmother Bell's striking looks and flirtatious manner.

For the most part, the children ran riot. Taken to school by the coachman Joe, who was Irish, they would screech 'Rule Britannia' and 'The Wearing of the Green' at the top of their voices. Alice never noticed whether they attended lessons or raced the carriages up the sidewalk without bonnets. She insisted only that Sunday be a day of rest. But her mother was horrified at this ill-disciplined brood.

Todd was too preoccupied to instil much discipline. When he did see the children he indulged them. Every week when he was in town, Todd would visit the mail-steamer as it arrived off Glenelg. On one trip he took his two sons with him. On his return, Alice asked what he had done with her boys. 'What boys?' Todd replied, vaguely. Then, more teasingly, 'No one would want such boys except you, they will be returned.' Not long after, a red mail cart deposited the refugees at the door. Todd noted down these few occasions of levity, when he had managed to make his wife laugh again. He knew Alice was often unhappy, but he neither had the time nor the courage to work out what was wrong, and assumed she was merely finding it difficult coping with Adelaide society and all the carping about his job.

Alice's post-natal depression must have exacerbated her melancholia. While Todd buried himself in his red earth, Alice yearned for the cool climate of Britain, where her family knew their place and everything was ordered. Here her husband was always either up or down, in or out. She never knew when she would get the next invitation to the Governor's residence. But she still defended Todd to the hilt.

The preparations for the trip to the Roper in 1871, however, brought Alice and Charles closer again. Alice made Todd a new set of underwear and handkerchiefs, and embroidered them with his initials; she also lined his waistcoats with green silk and polished his belt. Many of the letters the couple sent each other while Todd was working on the lines were burnt by Lorna, but the letters they wrote while Todd was on the Roper still remain.

Back in England, I read Todd's Roper diaries, flicked through the letters between the couple, and looked at Todd's notes to his children about the man-eating crocodiles. If he could make it up north, so would we. Ed was particularly intrigued by the sound of the fishing on the Roper for twenty-pound barramundi. So, six months after our last trip, we set out from Britain again. This time we were aiming to get to the Northern Territory in the middle of the dry season.

19

·———— ————·

The Stuart Highway

In 1871, Charles Todd wrote down a list of what to take:

4 trousers,
3 dozen collars,
3 alpaca coats,
2 prs boots,
8 prs drawers,
3 waistcoats,
3 pyjamas,
3 scarves,
10 towels,
4 Crimean shirts,
5 jerseys,
1 clothes brush,
1 pr slippers,
3 puggarees,
12 prs of socks,
1 belt,

1 green scarf,
9 handkerchiefs.

He had been planning to take his cabbage-tree hat, a canvas pudding basin that barely covered the neck. But when he went to the Governor to wish him goodbye, Sir James Fergusson, who had spent years in the West Indies, insisted he take a solar topee instead. 'Life in the tropics is a different game,' the Governor told him, and showed him how to wind a puggaree round the helmet so it fell down over the back as shade for the neck. Todd rather liked his new white headgear. The helmet added nearly a foot to his small stature. He paraded round the house in full regalia for the benefit of the children, and sent them pictures of him posing in his white uniform. His daughter Lizzie later wrote to him: 'I don't think the photo of you in your bush clothes is very flattering.' He also took with him his favourite book, *The Ancient History of the Abdication of Kings*.

This time Ed and I decided to be slightly more economical with our luggage. I found the list from our previous visit in our diary:

1 pr sandals,
1 pr trainers,
3 prs socks,
5 prs knickers,
1 pr jeans,
2 prs shorts,
3 dresses,
5 T-shirts,
2 cardigans,
1 belt,
1 bikini,
sunglasses,
sun cream – five strengths,
contact lenses,

cleanser,
moisturiser,
earcleaners,
camera,
watch,
5 packets jellybeans.

On that occasion, of course, the airline had lost my suitcase and someone, somewhere, was eating my jellybeans, or worse, wearing my dresses
and finishing off my moisturiser. This time, I pared my provisions down,
added walking boots, and insisted on carrying only hand luggage. Ed just
packed his Akubra hat, a pair of shorts and seven T-shirts.

Finding the disposable paper knickers Pat had bought for me made me
smile. I'd been shocked when I had learnt that Pat had died three months
before this trip, and I missed her. She had told us she was ill, but, typically,
had underplayed it. In fact, she had had cancer.

Her letters back to England barely mentioned her illness; instead they
were crammed with advice on the book. 'Did you know that Todd's
favourite food was rice pudding? He liked to take it cold with him into the
bush. Did you realise Alice kept a butterfly garden?' Pat was the last link to
Todd, the only one who had listened to all Aunt Lorna's tales and kept the
mementoes together at the bottom of a bookcase. She pretended to be a
tough old nut. 'Why do you want to write a book about Todd, haven't you
got better things to do?' she would ask us. 'When are you going to settle
down and start a family?' I wish I'd known her for longer. She was such an
extraordinary woman, running her property on her own, and so feisty. I
could still picture her rounding up the cattle with Todd's whip.

I remember her showing us how to make the bushman's staple,
'damper', in her back garden one afternoon in case we ran out of food. She
mixed up a potion of water and flour, and then wrapped the dough round a
stick and stuck it on the barbecue. This white, gooey mess puffed up and
swelled until it looked like a giant scone. She went into the kitchen and

came out bearing thick cream and Cambridgeshire strawberry jam to spread over our concoctions. They were quite delicious. 'There won't be any cream in the outback,' she warned us, as we lay on the grass trying to digest the equivalent of several Cornish teas. It would be difficult returning to Adelaide knowing that we wouldn't see her again.

There was a sweltering heatwave the day we drove to Heathrow. 'Thank God we're leaving this heat behind; it'll be winter in Adelaide,' I said, asking the steward for more blankets. 'You sweet?' he said. This obviously meant, 'Do you want an ice cream?' I explained to Ed, feeling my antipodean roots stirring. 'I think he's just asking if you're comfortable,' Ed replied, 'and you've forgotten, it'll be forty degrees in the tropics.'

At least it was cold when we arrived at Adelaide airport. Even being douched with disinfectant by the quarantine officers didn't dampen our enthusiasm. We went to see Pat's brother, Bob Fisher, a Supreme Court judge. The sign on his office door read 'Nunc Dimittis' and inside the walls were lined with dusty legal tomes. Bob, when we discovered him reading in the shadows in an old brown chair, was as small and lithe as Pat. We chatted about his sister and her property, which had now been sold, and he explained that he had once owned Todd's revolver. But Bob was much more excited by the other side of the family, the Fishers, who were even older Adelaide stock than the Todds. 'The second Fisher was the man who made the money, he came over in 1838 aged three, and never looked back, fascinating character, so much drive.'

There was only one branch of the family left in Australia with the surname of Todd. Descended from Hedley, Alice and Charles's third child, Barry Todd lived in the suburbs of Sydney and had three grown-up sons. Barry had told me over the telephone to England that he thought Charles Todd was a far greater hero than Scott of the Antarctic, just less appreciated because he was an Aussie and 'a hillbilly South Australian to boot'. Ed and I decided to fly to Sydney to meet them. After all, this was our third visit to Australia and we hadn't even gone to Botany Bay. Barry was on holiday in

England, but Julian, his eldest son, sounded chatty on the phone, so we climbed on a plane, promising to meet him at his parents' house where he was staying with his three children.

Julian was far better looking than Charles, broader and more at ease in his sweatpants and singlet. Fittingly, he worked for a telephone company. 'Have you had any tea?' he inquired, as the children screeched round the house. 'There's a great drive-in McDonald's round the corner, it'll shut the kids up.'

We piled into his people carrier, and, after helping spoon Chicken McNuggets into my new-found cousins, I quizzed him on the subject of Todd. 'What about Todd's will?' I asked. 'Do you know where the original is?' He didn't. Had he ever been to Alice?

'Only once, for the regatta, but it got rained off. Freak weather, they said, so we went on to Ayers Rock.' I had thought that our family were the wimpy ones, escaping back to England, unable to stomach life on the frontier. But Julian made me realise that suburban Sydney is just as far from the outback as London.

We spent three days in Sydney before heading back to Alice and the cold/hot heat of winter in the centre. Peeling off the sweaters, my arms were still crisp from the last trip. The camper-van was waiting for us at Ian and Diane's. 'No floods this time, ' Ian shouted out. 'It's as dry as a nun's nasty.' He laughed.

Ian was soon telling us about his new trainee, who had arrived from Melbourne with a ponytail, shorts down to his ankles and his cap on backwards. Finding Ian washing the Land Rover, he mistook his new boss for the cleaner and asked where 'the fat slob' was. Ian had straightened himself up to his five foot seven, and said, 'If you want the job you'll be back here in an hour with your hair shaved and a better attitude.' To Ian's amazement, the apparition returned wearing long white socks and smart new khaki shorts, bush-ready, as the Australians call it.

'You're too fat,' Ian told him. So the poor boy was made to cycle to work every day, and run during the lunch hour. 'He loved it. In fact I was trusting him to take over the business,' Ian said. 'Then we came across a decapitated body in a jeep out on a recce one day. The jeep must have hit a ditch and ricocheted into the valley. As a soldier I've seen a few body parts in my life, but my number two never recovered from picking up the head, so I'm on my own again.'

Even Ian was finding the outback tough. 'You don't get anal inspections here, like we had to suffer in the army, but all the bureaucrats buggering around get me down. There's politics even in Alice. The tourists don't want quality, just a quick gawp at a kangaroo and the operators only want to turn a fast buck. Alice has lost its innocence. We're heading off to Perth.'

To prove his point, he took us to a political meeting. All the town's leading tourist lights had gathered to greet one of the new Northern Territory ministers over ham rolls. They assembled at a new desert park, and Ian delighted in pointing out the local nymphomaniac. This formidable lady was horrified at government plans to improve the road network. 'Tarmac and signposts will mean the end of our tour companies,' she said. 'The tourists would be able to drive themselves to the sites.'

A property owner shouted out from the back: 'You're so goddamn selfish. Haven't you thought who else might want better roads?' There was silence. Only a question about the Aborigines united the audience once more. A respected local flying doctor said, between mouthfuls of curried egg, 'Half my work is looking after drunken abos, who've been beaten up by their husbands. What are you going to do about it?' The minister cracked his knuckles slowly, shook a lot of hands with a firm Territorian handshake, and got back on his plane to Darwin. The man next to me was muttering, 'If you find uranium under a rock that the abo hasn't been near for thousands of years, it soon becomes sacred coon country.' He saw he'd caught my attention and smiled. 'We'll sort the vermin out ourselves, if the politicians won't.'

Feeling embarrassed, I spent the afternoon trying to arrange a few days with some Aborigines. But their guardians were more supercilious and protective than the Buckingham Palace press office. The idea of doing a little snake-hunting and lizard-eating with them was as ludicrous as suggesting going hunting with the Queen. We should have rung six months ago to arrange an audience. None of the Aboriginal leaders would be prepared to talk to anyone straight off the bitumen. We needed to be positively vetted to make sure we had the right sort of prejudices.

'They're not just here as exhibits, you know,' one woman said. 'I've tried talking to the Aborigines in the mall, but it's very difficult,' I explained. She went ballistic. 'I hope you didn't take photographs?' No, we just wanted a chat. 'A chat? Why should these people want to converse with you?' 'Look, I know the statistics, I know that Aborigines are imprisoned at twenty times the rate of white men, I just want to hear whether they have any stories about the telegraph line,' I pleaded. I could hear her earrings shaking from side to side. Wearily, we did what she insisted and booked ourselves on an expensive three-day course of basket-weaving and spear-throwing taught by Aborigines near the town of Katherine, another week's drive up the line. We chose the course from a special 'Dreamtime Brochure' and Ed spent the next few days practising throwing sticks.

Ian was toying with the idea of coming with us to Darwin, but Diane reminded him that she was seven months pregnant and didn't trust us to deliver the baby if she went into labour over the ruts to the Roper River. The next day, Ed, having tried the sleeping compartment in the camper-van, announced that it would barely fit one reclining adult, and certainly not four. He sloped off for a last espresso, so I set out for the supermarket alone. I remembered the first law in outback cuisine: however murderously scorching the day, you can only ever buy hot, fried food. We would soon have our fill of pineapple and sweetcorn fritters and battered sausages. What we needed were lettuces, tomatoes, olives, mozzarella and feta in our mini-fridge so we didn't get osteoporosis or scurvy.

By the time I'd finished loading everything into the back of the camper, it was early afternoon. This time the trip should be easy, since it was bitumen all the way to the exit for the Roper. We set off up the highway to

Central Mount Stuart, swaying in the wake of the road-trains. The tarmac drew a harsher, hotter line through the scrub than the red dirt roads of the south. Two hours later, we could see the hump in the distance. We stopped off for a drink at Ti Tree to reacquaint ourselves with the outback pub. The stickers were still the same. 'There's no such thing as a rape; women run faster with their skirts up and their knickers down', and 'The music of a working 4.2 will always piss off a Subaru. The growling of a Two Five Three is enough to make any feral flee.'

The barman glared at our English knees with the same distaste that I remembered from last time. The Centralians must be the least fashion-conscious people in the world with their cripplingly tight shorts, their moneybelts for cod pieces and their maroon singlets, but that doesn't stop them staring. I bet Ed £100 he wouldn't be able to eat one of the meat pies wilting in a cage on the counter. 'Do you want me to be ill again?' he asked in amazement as a trucker took a huge bite out of one to find a greying, shrunken meatball at its core.

We headed for Central Mount Stuart, the geographical heart of Australia. The solitary hump had become three or four bumps, and there was no track towards them, only barbed wire along the roadside. Eventually we found a gate with a 'Dangerous Mining Area, Keep Out' sign, and headed through on to a sandy track, bumping over the spinifex towards our goal. By the time we got to the base of the range, the car looked as if it had been attacked by a thousand sharp-nailed lovers and it was getting dark. So we took out our folding chairs and table, and made our salad under a desert oak.

The sun hit the peaks in a Mills & Boon orgy of red and orange just as we were finishing our coffee. But there was nothing romantic about our bed; the one-foot-high sweaty hole at the top of the van wouldn't accommodate an old-fashioned embrace. To sleep at all, we would have to get undressed, wriggle almost naked into the compartment and then try to drag our sleeping bags over our feet, keeping our elbows tucked in and our faces turned to the side to catch the occasional puff of fresh air from two ventilators. A single mosquito caused havoc. Rolling over was impossible, and it would have been easier not to drink all day than risk going to the loo at night. By the time we had managed to wriggle out of the van, an adder was sure to have spotted our white feet. I was convinced that our childhood camper-van had been more luxurious. 'You were just smaller,' said Ed, who finally went to sleep wrapped round the steering wheel. All we needed, I decided, as dawn ended our night of contortions, was a brisk walk up the hill. 'What about a shower?' Ed complained.

Stuart, Giles, Todd and the many other explorers who made the trip up the mountain never mentioned the climb. Nor were there any signs to help us as we headed for the top of what we assumed was the tallest peak to see if we could locate the site of the bottle that Stuart had buried on his journey across the interior. Stuart wrote that he had climbed the mountain and built a large cone of stones at the top. In the centre he had placed a pole with a British flag nailed to it. 'Near the top of the cone I placed a small bottle, in which there is a slip of paper, with our signatures to it,' he wrote. 'We then gave three hearty cheers for the flag, the emblem of civil and religious liberty.'

For three hours we struggled through the scrub until our legs were bleeding. By midday we had left the trees miles behind and were shaded only by our hats. We were still arguing over which was the true peak, but the views were phenomenal as we hovered on our tightrope above the Stuart Highway. We had finally reached the continent's centre.

The bottle had long since gone, so we had nothing to do except watch a distant kangaroo while Ed explained that this species had the largest genitalia in the world. 'Bigger than an elephant's?' 'Far bigger,' Ed said solemnly. 'They're prodigious.'

The scramble down was even more fraught. 'This is unacceptable,' said Ed as we reached the van, only to find that the mini-fridge had failed and the humous had leaked into the feta and capsicums. 'Unacceptable' is Ed's most chilling form of rebuke. 'I'm not climbing any more mountains. The whole point about desert is that it's flat. And while we're here, I want to eat steak not some New Age crystal-wearer's lunch,' he said. We didn't have any choice. It was dry Frosties for tea, dinner and breakfast.

The Stuart Highway wasn't proving as relaxing as we'd thought. The road-trains were like whales; they commanded respect and distance. It was the Subaru drivers from Alice who crawled up your bumper whom you wanted to throttle with their tow-ropes. 'If they break down, we're not helping,' Ed said, as he swapped another V-sign.

By the time we reached Barrow Creek at noon the next day, we'd settled into the highway's groove and were the only ones left on the road. Barrow Creek was the site of the first Aboriginal attack on a telegraph station. Inside, the pub had the obligatory unseasonal Santa Claus coming through the roof, and barman with shaved crown and long pigtail. A hatch in the wall served hot-dogs and beer to the local Aborigines between noon and three p.m. 'They can have six cans a day,' the barman said.

We drifted out past the truckers sweltering in their lumberjack shirts to the old repeater station, so viciously fought over for its first fifty years, now almost forgotten. It stood alone, two stone sheds and a house. The supervisor of the local Aboriginal settlement lived in a luxury caravan nearby, surrounded by three jeeps. Inside was a powerful computer and a white woman doing the accounts. 'Dad's out and Mum doesn't want to talk; she's counting out the cheques,' said the daughter.

The barman got on well with the Aborigines. Mike explained that he was from Adelaide and his first job had been teaching English on an Aboriginal settlement. 'It was gutting, mate,' he admitted to Ed. 'The dogshit was what got to me, and the dogs howling at night. The kids were lovely – triffic, but I felt like a prison warden. Here it's easier. It's depressing watching them queuing up every morning, but at least I'm giving them what they want. They get through fifty ice creams and forty-five meat pies in an hour, and if I buy in some nail varnish they snap that up too, some for sniffing, some for painting. But I don't sell perfume, or they drink it.' The Aborigines obviously liked him, taking him with them when they went off hunting emu or kangaroo in their Land Cruisers.

Mike took us outside to introduce us to an Aboriginal woman called

Nina, and suggested we buy her a steak sandwich. Nina spoke a little English and talked warily, staring at my wedding ring and feeding bits of tomato to the tame cockatoos. She pointed to the hollow where she lived temporarily. She had five children and a house, 'But it's shit,' she said, offering to sell us a painting of a witchetty grub for A$50.

Disconcertingly, she was wearing the uniform of a Notting Hill trustafarian. In her hair was a spangly clip, and on her feet were a pair of flip-flops with plastic flowers. Her dress was pale pink. But her skin was pitted and her curls were matted. She looked like a hibernating bulb rather than an English rose, and her handouts came from the state, not Daddy – 'sit-down money', as the locals called it. We spent the day losing at cards to them, listening to a fusillade of shrill abuse from one wife to her husband, and buying rounds of ice cream.

20

.. _ _ _ _ _ _ _ _

Pole Fanciers

The next night we knew we would be among our own tribe – the pole fanciers, a group so small that it could congregate in the small town of Tennant Creek without anyone getting their singlets in a twist. We'd heard rumours about the reunion while in Adelaide, and we'd seen the fliers on a bar counter in Alice. 'Come and join us at the Tennant Creek repeater station for the 125th anniversary of the Trans-Australian Telegraph, bush tucker and beers, tickets A\$20. Lots of fantastic prizes including a signed Charles Todd insulator.' Just as we were leaving Barrow Creek, we saw someone had scrawled over the pink and green invites in large letters, 'GET TO MEET TODD'S GRANDDAUGHTER ALICE IN A RARE QUESTION AND ANSWER SESSION – SHE'LL SPILL THE BEANS ABOUT EVERYTHING YOU WANT TO KNOW ABOUT TELEGRAPH TODD.' 'Rare,' I spluttered. 'I would have been dead for thirty years if I was his granddaughter. This is too humiliating. I'm not standing in front of a group of history boffins being quizzed about the size of his toes.' (The second one on his right foot had been chopped off after a horse had trodden on it, according to Pat.) 'Anyway, who told them I was coming?'

'They heard it on the telegraph wires,' Ed laughed. 'Actually, I rang the number and explained that we'd like some tickets for his grandchild. I didn't think they'd get so excited.'

'But you've missed off several greats, they'll be expecting a centenarian. That's probably why they're so excited. It's so embarrassing,' I said.

'Don't be pathetic. This should be the highlight of your adult life, your starter for ten to boast about your expertise and swap stories about your favourite pole-hugging experiences,' Ed replied. 'I can't wait.' He was choking so much he couldn't start the car. I spent the next few hours learning my dates.

Tennant Creek spilled off the bitumen ahead and I still couldn't remember Todd's younger brother's name. 'Don't worry, there'll probably only be twenty fanatics there,' Ed said. A welcoming committee was standing inside our motel foyer comprising the Northern Territory National Trust Council, a mixture of petite housewives with matching suntans, a blacksmith from Alice Springs, an anthropologist and a scholarly looking professor, all meeting in Tennant Creek for their biannual knees-up. The motel owner asked if we wanted air con, before slipping in: 'You know, my great-grandfather was on that first telegraph trip too, you've probably heard about him, his name was Smith.' 'We're very glad to meet you,' said the chairman looking perplexed at my sprightly gait. 'You'll probably want a wash down,' he added, trying not to allude to our five-day-old sweat. 'It's like meeting Shakespeare's grandchild,' said one woman, patting me on the arm. 'Kick-off is at six p.m. Don't be late or we'll miss the food. They're expecting three hundred.'

We'd been booked in under the name of Todd, but all I could think of was the crowds, as I forced Ed to concentrate on choosing between my clean jeans and a dress. We clambered into the minibus together and, as we bumped down the gravel track, I voiced my fears to my neighbour. 'Don't worry, dear, no one is expecting too much. Tennant Creek will accept any excuse for a party.'

The repeater station was festooned with fairy lights. A truck sold cold beers, the kitchen doled out authentic 'pioneer tucker' – corned beef and damper bread – and in the centre of the courtyard was a large stage and microphones before a camp fire. Our National Trust friends kept a close eye on us, bagging a table under the tree so we could all reminisce about our favourite angle poles before the festivities began in earnest. They elbowed us to the front of the food queue and had smuggled in some wine. The amateur dramatics society was wildly applauded for re-enacting the planting of the first pole. But I couldn't be drawn away from the bric-a-brac table, where a line of dolls in various outfits was laid out. There was Wonderwoman in her cape, Mother Mary, Florence Nightingale, an unmistakable Joan Collins, the Queen with her corgis and Alice in Wonderland. I bought three.

'I'm not carrying those hideous things across Australia. You're going to be so embarrassed when you get back to England,' Ed said when I showed him my purchases.

'But you don't understand, they're insulators in costume', I replied. I lifted the petticoats on one doll to reveal a china insulating bottom. 'I'm going to arrange them along the dashboard.'

'But I thought you wanted them to remain buried in the desert for posterity,' Ed reminded me.

A local journalist led me away for an interview behind the stage, so I didn't need to answer. After being quizzed for half an hour about why my family had left Australia, and having posed with the local MP, I was introduced to Kathleen, who had been brought up at Tennant Creek repeater station. Her mother, an Aborigine, had been the cook; her father had been a telegraph man. I explained why I'd been called Alice and she laughed. 'My half-brother was called Telegraph, so you're lucky.' What had it been like growing up on a repeater station? She shrugged, leaving me to discover that as a half-Aborigine she'd been taken away to Alice Springs when she was seven and only managed to make her way back to her mother after eight years. She'd forgotten her father's name. Had the stringing of the line across Australia ruined her land? Of course it had. I didn't need to ask.

I returned to the middle of an auction. Ed was already bidding furiously for something that looked like an old envelope, but was actually a first edition set of Australian stamps from twenty-five years ago, commemorating the centenary of the overland crossing. I was touched. I was far more grateful five minutes later when the primary school 'folk dancers' had finished their rendition of 'I'm a little teapot', and a ranger from Barrow Creek station bounded on stage. 'I'm all sure you know what we're here to celebrate, boys and girls?' A few well-primed children put their hands up. 'Yes, that's right, it's to mark the 125th anniversary of the laying of the telegraph poles. And do you know how many poles were needed to go all the way across Australia?' Forty thousand, said a small girl. The children seemed to have won brownie badges on the subject. The ranger continued: 'And without this repeater station, Tennant Creek would never have existed, would it? No swimming pool, no Magnum ice creams, just desert. We can barely imagine what those men must have suffered on that first heroic expedition. So let's give a round of applause to Charles Todd's grandchildren, Alice and Ed.'

Ed climbed on to the stage. I've been to press conferences with prime ministers and have had no qualms about asking Serbian militiamen about their human rights records, so I can't pretend to be shy. But faced by all this sudden interest in a project that until now had been utterly personal, I couldn't say a word. I hid at the back of the crowd.

Ed, realising that his wife had done a runner, tapped the microphone and began. He was brilliant. Those days spent moaning about his bronchitis from the back of the car, the hours wasted having to discuss engine sizes with the truckers and the scars from shaving in the wing mirror were momentarily forgotten. He was flawless, his very English accent ringing round the courtyard. He raved about the beauty of the outback, the bravery of the telegraph men, the heat that could buckle an axe. He even took a few questions.

He was followed by 'The Bloodwood Bushband'; three enthusiastic, long-haired, middle-aged crooners wearing bandanas. They started playing

'Waltzing Matilda', and our friends from the National Trust asked us to dance. David was the obvious leader, a calm, collected professor of history in Darwin. One woman came from Katherine, the largest town up the road. Her husband had moved there because he was an obsessive fisherman. 'But it's desert,' we said. 'Not in the rainy season. Then we get some of the best barramundi fishing in the world. My husband just sits out all day in the garden, reeling them in.' Her friend from Pine Creek interrupted. 'In the rains you can't drink the water, or you go mad. It's called the suicide season, the heat just builds and builds and you want to run into the street naked, screaming.'

Their hobby threw them together. They spent eight hours the next day discussing forward planning. 'There aren't very many old buildings left in the Northern Territory; those that weren't taken down by Cyclone Tracy have been eaten by termites,' they explained over a finger buffet of authentic modern tucker: tinned beetroot, tinned potato salad, tinned peach slices and fried chicken wings.

We didn't want to leave Tennant Creek. It may have been a one-road, one-traffic-light town where a child had died when he strayed into the heat outside his backyard, but someone had lovingly planted roses along the highway and we'd miss the school athletics races the next week. 'We'll have champagne,' said the motel owner.

An identical camper-van drew up at the petrol station just as we were leaving. 'Snap', said the girl jumping out, in an unmistakable English-public-school voice. Four immaculate friends followed her, their squeaky clean ponytails swinging from side to side. Lavinia, Clarissa, Isabella, Lucinda and Camilla (or Lavie, Clarrie, Isie, Lucy and Millie) were on their gap year. We wanted to know where they all slept. 'Three on top, two under the seats', they chorused. Their van had a teddy-bear mascot dressed in Chelsea strip hanging from the mirror. What about the heat? 'Oh, it's nothing compared to India, Thailand, Borneo, Bali and Sumatra,' they chorused. Lavinia was wrapped in a scarf, looking coy. 'Lav's got a love-bite from a trucker,' they laughed, 'and

Lucinda got a tattoo in Sumatra, she didn't even have an anaesthetic.' These were the inheritors of the Victorian-lady-traveller tradition, to which Alice, saddled with five children, had never belonged.

As we headed further north, the scrub got thicker and a few gangly trees appeared. The National Trust had told us where the memorial to Todd lay, and we stopped at the lay-by to read the obelisk. Beside it sat a heap of dog biscuits covered in ants, and a single metal pole. The road-trains spewed us with dust but we were happy to pay homage to Todd.

A few minutes later we were back on the road and arguing. 'Did you see that sign, "We like our lizards frilled not grilled?"' I said.

Ed laughed. 'Don't be ridiculous, it was: "We serve our lunches fried or grilled."

We reversed back along the road until I had been proved right. We had now entered the real Northern Territory. This was holster belt and pistol territory, a million miles away from the manicured south. The Top Enders still pride themselves on straight talking and hard riding. The sign was the Territorian way of saying, 'Please don't start any bushfires or you will endanger our wildlife.' When they had installed the first traffic lights in Darwin, the notices had been even sharper: 'Look, these are not bloody Christmas decorations. If you don't do what the buggers tell you, you'll be booked.'

Turning off for Daly Waters, where one of the original stockroutes met the path of the telegraph, we headed for the oldest pub in the Northern Territory. Three miles later we bumped into town, past the cemetery, the disused telephone box and an old aircraft hangar. Todd established a base at Daly Waters, drovers used it as a camp-site and explorers would come to stock up on water. Amy Johnson had landed her plane here and the army had requisitioned the town as its final redoubt against the Japanese. Now its population seemed to have dwindled to three: a young boy behind the bar, a man slumped on the stool and a Telstra engineer who was here to mend cables.

'Do you know if there are any remnants of the old telegraph wire?' we inquired.

'The history's on the board, luv,' grunted the man on the stool. We looked up at three grease-stained, dart-pocked signs with pictures of cattle drovers, telegraph men and the army.

'This pub can't have been open since 1862 – no one came to the Northern Territory until ten years later,' I pointed out.

'Don't come the raw prawn with me, girl. Just coz you all live in some bleedin' medieval castle, doesn't mean you have to be a wowser. This is the most authentic bloody boozer in Australia. Do you want a stubbie or don't you?'

We took the beers and shut up.

But in a country this thinly populated, people can't hold grudges, and we were soon best friends with the bar-stool man and spent the rest of the evening drinking white wine mixed with 7Up. 'How the coolies drink it, mate,' he said, and started reminiscing about Vietnam. 'Yep, we were there, mate. Didn't think us Aussies got our feet wet. Well, we were slaughtered, just the same as the septic tanks.'

The septics?

'Yeah, the Yanks. Came back, hung up a few beads and set up a hippie hang-out up in Darwin with the wife. Only she legged it with our lawyer. Nearly bankrupted me. Then Shiklone Tracee wiped me out. Now I'm in this Pommy's towel of a desert, with a shonky shitbox of a car, and I'm skint.' And after my fifth glass of Sparkling Up Wine I was seeing double.

The next evening we opted for the safe-sounding Mataranka Homestead, eighty miles up the tarmac by the turn-off to the Roper River. 'The crystal-clear thermal pool, in a pocket of rainforest, is a great place to wind down after a hot day on the road – though it can get crowded,' said our outback guide. We even splashed out on a portaroom, 'semi-carpeted'. The caravans leered back at our headlights like a group of indolent cows. There were hundreds of them, herded against the fence. Campers were queuing up outside the communal showers, soapsuds overflowed on to the brown grass. We squeezed our van into the last space available next to the loos and went in search of our room.

'Who's flushed tampons down the toilets?' boomed the receptionist as we waited for our keys. 'You think we're going to fish them out?' She turned her attention to us: 'Do you want aerobics in the morning or the nature walk? Towels will cost you an extra $3 each, and don't try taking them to the thermal springs. Supermarket's on your right, cafeteria on your left, it's fried chicken or steak tonight. There's a disco on the patio.' Aussies had come for miles for this hot hole in the ground. Most of them were pensioners on walkabout, curious about the continent that they'd shied away from during their working life. Now they were following the highways like songlines across the central plains, convinced that somewhere they would discover the real Australia. And they were returning, without realising it, to the ways of the original Australians.

The next morning, the thermal springs were already covered with a thin layer of suntan oil and a thick layer of pink people. We turned off the tarmac and embraced the rutted, corrugated dirt-track to the Roper River with relief. Where we were going, the caravans couldn't follow.

21

.. ___ .____

The Roper Diaries

When Todd embarked on the *Omeo*, the government gave strict instructions that he should be dropped with his men, fifteen wagons and eighty-six horses at the mouth of the Roper, and the boat should then continue to Darwin. Todd knew this would condemn him to failure before he had even started. If his men had to walk a hundred miles into the interior along the Roper, laden with provisions in the downpour, they might never reach the base camp. So he immediately collared Captain Calder and pleaded with him to take the *Omeo*, a vast barque-rigged sail steamer, up the narrow, unnavigated river into the jungle, a feat comparable to taking a destroyer up the Thames to Henley. Calder consulted the owners of the barque and the message came back. 'Follow terms of charter. If any deviation Government must cover all risks . . .'

The *Omeo* was packed not only with insulating wire, poles and cattle, but new settlers bound for Darwin, a rash of gold prospectors and a criminal sentenced to ten years in jail as a suspected cattle rustler. Andrew Hume had talked his way out of jail by promising that only he knew where the relics of Ludwig Leichhardt's doomed expedition to the interior lay. The Sydney government had insisted that Todd take him along, but all the men on the boat were convinced he was an impostor, and indeed he disappeared into the outback before the floods were over.

Patterson had prepared his final report for Todd:

> I put forward no excuses . . . for want of my success, because such excuses would only infer self-accusation, and from such upbraidings, I am entirely free . . . The expedition has not suffered from misfortune so much as mistakes . . . I am suffering from the imputation of disasters which were inevitable under the circumstances, but which would undoubtedly have been averted, had I been left unfettered in my action, and been allowed to carry out your original idea of making the Roper the sole or main base of operations.

Todd barely recognised the shrunken Patterson when he boarded the *Omeo*. He was even more startled when Patterson immediately insisted on resigning. The last thing he needed was his subordinate spreading despondency down in Adelaide while he was trying to finish the line. So Todd refused to accept the resignation, explaining that he was just the facilitator. Patterson should use him like an axe, to help cut through the problems.

Patterson was not so easily mollified, and insisted that Todd spell out their positions in writing. Todd set out a three-page letter exonerating Patterson for previous failings, but making it clear that he now held overall responsibility as superintendent. 'I propose that you should still act as chief officer of the expedition and have principal charge under me of the parties employed,' he wrote. Patterson wasn't satisfied, calling Todd's reply 'evasive and foreign'.

Todd had had enough. 'I shall be glad to know whether you are prepared to recognise my authority and retain command of the expedition?' he wrote. Little enjoyed acting as go-between in this bizarre exchange between cabins.

The two men eventually shook hands, and Patterson filled Todd in on his meagre achievements. Patterson's deputy, Rutt, was taking his team south from the King River. McLachlan's team was working south from the Elsey, and no one had any idea where the team furthermost south, under Burton, had reached. Todd soon discovered that the situation was even gloomier. Rutt had barely poled twenty miles when the floods rose. He was now camping on the highest hummock, which he had named Providence Knoll, having waded waist-high through the lakes with the stores and equipment, and was waiting for the waters to recede. McLachlan's team was sheltering on another hill, and Burton had only erected ten miles of line before the rains began. The men were now in danger of starving. The rainy season had surpassed itself. One of Rutt's team wrote, 'The atmosphere in the tents was so heavy, that we had to go outside, despite the torrential downpour, so we could breathe.'

There was nothing Todd could do until he could get the *Omeo* up the Roper. The boat had now been stuck on Maria Island, just off the coast, for several days, with no sign of the *Young Australian* that could have towed it upstream. Todd begged the captain to risk navigating the tricky Roper on his own. Captain Calder pointed out that the *Omeo* was an ocean-going steamer, and that a sandbar ran across the mouth of the river. But he offered to try his luck if the government would pay full insurance (£20,000) if the boat ran aground, and if he could be excused from continuing on to Darwin, as he needed to be back home in Adelaide in three months. Todd, exceeding his authority, agreed.

So Captain Calder rowed out to sound the bar at half-flood tide and promised to hoist sail if he thought the ship could get across. For what seemed hours, the boat traversed the bar. Then the flag went up. The ship waited for full tide at four p.m. and slunk over the bar. With men taking regular soundings, they started to creep up the river. To both Todd and Calder's relief, they found that the bottom was sandy, not rocky, so at least the ship would not risk being holed if it grounded.

The ocean steamer must have looked a strange sight to the Aborigines as it slipped through the interior. Its masts stood out above the mangrove swamps and grassy plains. After three days, they had covered over seventy miles. They stopped to unload the restive horses, and the women asked to be allowed off to do some washing. An officer escorted them to shore and even put up a washing line. The women bathed their hands and faces, pounded their clothes, and were enjoying an impromptu picnic, when suddenly a group of Aborigines came up behind them. Three ladies immediately fainted, the others started wailing. The Aborigines were merely watching, but no one could get back on the boat fast enough.

Soon there were four ships on the river. Patterson's cutter, the *Larrakeeyah*, the sailing boat, the *Bengal*, Todd's barque the *Omeo*, and the *Young Australian* paddleboat now completing the fleet. The *Omeo* struck the bottom twice, but with three crews to help she was soon righted and the boats made a stately procession up the river into the jungle, their masts occasionally scraping the branches of trees. Aborigines ran by the side, imitating the paddle wheels with their arms. The ships docked side by side at the jetty as the men cheered. The passengers laughed that it was like being back at Richmond, there was so much traffic, and insisted that one day the Roper would be just like the Thames, and the camp of motley tents could become another miniature London.

Todd wrote to Alice: 'I think you will say it is a great feat to have brought up this new river, 85 miles from its mouth, the *Omeo, Bengal* and *Young Australian*. It is a great triumph for me over those who so strenuously opposed the Roper being used . . . the banks are thickly lined with trees . . . gums, paper bark, acacias, and a very fine ficus being the most conspicuous . . . The country is splendidly grassed.'

Two years later, Patterson was to write mendaciously in a memorandum,

That taking of the *Omeo* up an almost unexplored river . . .was entirely owing to me . . . the Superintendent of Telegraphs was credited by the press and the public with the success of this venture, and he quietly accepted without demur; whereas at the time he was violently opposed to the step, and was with the utmost difficulty persuaded by me to agree with it.

It was clear from the diaries of others taking part in the expedition, though, that the decision to take the ships up the Roper had been Todd's.

The crews assembled on the deck of the *Omeo* to hear Todd praise their work so far, and warn them that speed was now essential if the line was ever going to be finished. But as it was still raining, there was little anyone could do except unload the supplies. Worried that the men would lose heart, Todd took his officers into the saloon that night and over sherry gave an emotional speech setting out why the project had become so dear to him in the last few years. Even Patterson felt obliged to add at the end that Todd was a true friend and he appreciated his having come up to help.

The men, women and children had been forced to disembark because the *Omeo*, which had promised to take them to Darwin, was now returning to Adelaide. The hinterland soon took on the appearance of a thriving shanty town, with lanterns in the trees, tarpaulin draped over bushes, pots cooking on open fires and washing lines strung between telegraph poles. The children ran around half naked and learnt to swim with a rope tied round their middles. The port was christened Mason Town after the carpenter who erected a lone iron storage building amid the sprawl.

The *Bengal*'s Captain Sweet had promised to take photographs of the team, and Little, Patterson, Todd and Mitchell went off to change into clean bush-gear. Todd wore pristine white breeches, long dark-brown leather boots and a khaki shirt. He looked more like an advertisement for washing detergent than an explorer. Patterson had chosen dark fatigues, a stripy shirt and a straw hat. With an ethnic scarf wrapped round his waist, he could have been any public schoolboy traipsing through Thailand. With Little wearing two-tone brogues and Mitchell a string of Aboriginal beads, they made a queer bunch. When Sweet showed them the developed photographs they were appalled and insisted on starting again. This time Todd made sure he had taken off his glasses. He was standing on a knoll and looking to the distance with a suitably fervent glaze and his puggaree clasped to his hand, very much the team leader. Patterson was wearing his three-inch belt and a cap, having decided on the Afghan rebel look and Mitchell had discarded the beads.

Captain Calder, who admitted he'd rather enjoyed being part of the adventure, threw a farewell banquet 'with champagne for the officers and grog for the men', according to Patterson. Everyone went on deck to watch the extraordinary raw sunset, unlike anything they'd experienced in Adelaide. The *Omeo* steamed off the next day. Whether from the new lightness of the steamer, or the heaviness of the previous night's drinking, they managed to crash into a rock going round a bend, damaging the stanchions and the bulwarks. But no one, including Todd who was on board surveying the land, was deterred.

When Todd reached Maria Island, he gave the captain his dispatches and checked through his post, which included a letter from the new Premier, Henry Ayers, wishing him 'God's speed in your very arduous and important undertaking . . . not only for the honour and credit of our country but for the lustre which must be cast upon you for the conception and execution of so vast and difficult an enterprise.'

Todd also sent a letter to Alice, extolling the virtues of the Northern Territory and the plucky settlers he'd met on his journey up. 'At Bowen I had a glass of porter with the Deputy of Telegraphs and then we adjourned to a fruit shop where we invested in prime apples, plantains and each of a doll . . . Once a lizard ran up the inside of Mr Mitchell's trowser; he, of course, thought it was a snake and dropped his unmentionables with the greatest precipitation.' Todd raved about the new country. 'I wish you could see the river especially at sunrise and sunset, when the tints and reflections on the water are the most beautiful . . . last night, the first clear night we have had, I jumped for joy to see the seven stars of the Great Bear, a sight I had not seen for sixteen years. It was like seeing the cliffs of old England again, and awoke many reminiscence of days long since gone by.' He found life on ship mesmerising:

One night we had a terrific thunderstorm . . . The thunder is deafening, crashes following the lightning immediately. I was obliged to sleep below with the horses immediately overhead making an awful noise kicking, and . . . setting the little steamer going until you thought she would capsize. Then a horse plunged and got its forelegs into the sky light, and was only extricated with the greatest difficulty.

Heartened that the new government was evidently more supportive, Todd then commandeered the *Young Australian* to take him back up the Roper because he'd noticed that the horses they'd left on the riverbank were in danger from flooding. They got to them with hours to spare. The grooms and horses were huddled on a small hill, two hundred feet from the river, with most of their fodder washed away. They swam the animals to the boat, and headed for Mason Town.

The jetty had disappeared under the water. Todd had assumed that the rainy season was coming to an end, but the river had risen rapidly and the camp was now surrounded by new tributaries. The *Bengal* had become a floating hotel, charging exorbitant prices in the bar. The children were getting ill in the damp, mosquitoes were endemic and the bites were soon swollen and infected. Sleeping became impossible unless one first carefully burnt all mosquitoes with a candle as they landed on the net.

The restless Todd couldn't face being cooped up so he headed back down the Roper with the *Young Australian* to wait for more supplies, and to write his dispatches back to his subordinate, Cunningham, in Adelaide. 'My own health is wonderfully good, and I, by a semblance of cheerfulness I do not feel, am doing all I can to promote harmony of action and a more sanguine spirit among officers and men . . . This (the rain) dampens the spirit, many being sick of the climate and ill with fever and ague. I am compelled to supply them liberally with wines and spirits,' he wrote. To Alice, he added:

The mosquitoes invade us in such numbers immediately after sunset as to compel an instantaneous retreat to the friendly mosquito curtain in which we lie during the long hours of the night – if the rain will let us, but that often disappoints us of a night's rest on deck, coming on suddenly in a furious downpour. Your bed is flooded in no time unless you snatch up everything on the instant and rush down to the cabins.

He also admits, 'I am well excepting in the morning when I retch violently and have severe trembling fits till I can take some brandy and water. And after that, before breakfast I sometimes have to take a glass of sherry and quinine.' Todd sounded as though he had the classic symptoms of malaria.

Conditions upstream deteriorated every day. Soon the camp was on an island no more than thirty yards wide and 350 yards long. One passenger wrote: 'Really is most desolate. The collection of tents look like a number of discoloured pocket handkerchiefs spread out on a mud bank.' Some were too ill to leave their canvas. One young twenty-three-year-old, who had caught a fever coming down the line, died, provoking endless debate about whether to give him a sea or land burial.

Amusement for the healthy came from teasing two wealthy Englishmen who were on a grand tour of the new continent, and had been travelling to Darwin on the *Omeo*, complete with all the latest luxuries. They had two sets of travelling cutlery but hadn't got a sharp knife between them, and didn't know how to make damper. 'They had portmanteaus that can be turned into bedsteads, portmanteaus that can be turned into cooking machines, and sausage machines, and portmanteaus which, when all together, can be made into boats or gigs,' wrote one fellow camper. 'But luggage generally is a nuisance in this country. A suit of clothes that can be washed whenever it rains, a mosquito curtain, a knife and a piece of string are all that are absolutely necessary.'

Todd, meanwhile, was fuming on Maria Island, convinced he had been tricked by the scheming captain of the *Tararua*, who had taken precious weeks to deliver vital extra supplies. The man explained that he had arrived a month before, but, finding no one at the rendezvous, had assumed they had abandoned the idea of the Roper, so he steamed on up to Darwin, deciding to sell the horses there. Captain Douglas, the Resident, had had the quick wits to refuse to let him unload anything, insisting he went straight back to Maria Island.

The *Taranua* had brought nearly seventy more horses, but, to the stranded families' horror, the captain refused to go back to Darwin, and returned instead to Sydney. At least the floods were beginning to recede. Todd spent his time writing dispatches to the office about the rest of the line. He wanted the poles affected by white ants changed as quickly as possible, and all creeks to be crossed at right angles.

Patterson whiled away the long hours writing long letters to his wife – one of more than a hundred pages – incoherent with pain and anger, and insisted on naming all new rocks, boulders, knolls and islands after her. But his diary was his greatest solace. His outpourings were prolific. His thoughts on his wife were so intimate that in one letter he described himself as being 'defiled with sin'. When he returned he destroyed all the writings on her. He had been married to Elsey only for two years, almost all this time being spent apart, and had probably had little chance to enjoy the marriage.

Todd had little sympathy for this mawkish sentimentality, but then he was twenty years older and already a father of five children. Nor was he averse to naming a few tributaries and bluffs after himself. He wrote back to Alice, 'Much that you hear about Patterson is quite true. He was not fitted for this work and is too easily disheartened.' Another letter adds: 'Endeavour to counteract the dampening effect of Patterson's treatment of his men and the gloomy view which he takes of our prospects. I am always ready with a kind and cheering word or a timely joke. The consequence – I now hear that all officers and men swear by me.'

Even more smugly, he added, 'I seem to be always busy. Patterson is awfully lazy and selfish, does nothing but read, write to his wife, and hardly speaks to his men.' In his official letters, he made sure to praise Patterson, but to Alice he continued, 'He never works unless for some personal benefit and is dreadfully selfish. He has robbed me of my gun, taken all my saddle straps, wanted all my coffee . . . he is a great writer, always writing, always picking out smart terse antithetical sentences from books with the view to introducing them with effect in letters.'

In public, the two men kept a civil distance. Both were desperate to get out of the camp and into the interior to discover where the teams were, and whether they had survived the wet season. On 21 March, Patterson celebrated his twenty-eighth birthday. 'My birthday today,' he wrote. 'Have been considerably seedy and lying down the greater portion of the day. My heart sinks at the future.' Four days later, he left camp with four men and six packhorses to survey the mess. The project was now running six months late.

22

·· —— —— —— ·· —— —— ——

Dreamtime

As Ed and I turned off the bitumen towards the Roper River, we collided with a group of Vietnamese shrimpers. Having disentangled our bumpers, the men explained that their ship had broken down on the coast and they had been lent the pick-up truck to take their cargo to Darwin. None of them had a licence, but in payment for our mangled headlight they gave us a crate of iced prawns. 'They'll make good bait,' said Ed, who was looking forward to catching his first barramundi.

As the pink prawns gasped their last in the back, we could almost smell the sea across the sand. Our destination was the Roper Bar store where the telegraph men had set up their depot. It never did turn into a Richmond-on-Roper. The guidebook said there was one storehouse, a bungalow and a camping field for fishermen. We wanted to see if we could find any Victorian tent pegs.

For four hours we bounced along the red Borroloola road, sucking boiled sweets mixed with dust and debating whether we would dare to swim among the crocodiles. After a hundred miles we came to a drive, as grand as any country house, that swept up to a small shack and a petrol pump. 'Veronica's Store' was scrawled in felt tip on a wooden sign. Inside,

Veronica was shaking up the Hundreds and Thousands so they wouldn't get damp. 'Ve zell everything here,' she said with a 1930s Munich-beer-garden twang. 'You need ze pair of tights, I'ave three size. Pot-pourri und Pot Noodle. Vhat do you vant?' Veronica's arms were plaited with muscles, her bleached white hair stuck upright in the breathless air, her legs were taut mahogany and she was wearing well-polished army boots under her skimpy shorts.

We explained that we were looking for the telegraph depot. 'I not know what you talk about,' she said in an I-know-exactly-what-you-want-and-I'm-not-going-to-help-you voice.

'My great-great-grandfather came here to lay the telegraph poles,' I continued. The Valkyrie cocked her head. I resorted to O-level German.

She suddenly laughed. 'You don't need that. I just use the accent to scare away nosy parkers. Too many tourists and it gets crowded out here. I'll see if my boyfriend's got the boat.'

'So the river's still navigable?' we asked.

'Bigger than the Thames, not as big as the Rhine, or so my father, who's seen both, told me.'

Veronica's father, it transpired, was a legendary six-foot-six Aryan crocodile hunter who dressed in a loincloth to swim among his prey. 'He'd hunt them with a canoe and harpoon them. They may have gone under for an hour but as soon as they surfaced he'd stab them, avoiding the underbelly because it makes the best bags and shoes, and drag them back to shore.' Veronica's brother still water-skis when he comes to visit. 'Uses their heads as jumps,' Veronica smiled. 'A huge croc recently charged us when we were towing a car out of the river, got it between the eyes. But we're not really allowed to kill them anymore. We just trap them and sell them on to the croc farms for people like you to ogle at.'

Her boyfriend appeared in the doorway covered in blood. 'Don't worry, mate, just been gutting dinner,' he said. 'Veronica mentioned you were related to Todd. I'm a Telstra man myself. How can I help?' Glyn had left Hertfordshire for swinging Carnaby Street in the 1960s. From there he'd made his way round the world, following the Grand Prix tour before ending up in Adelaide. But it had taken another leap of faith to get him to the Roper. 'I was a happily married man with three kids until I came here,' he explained as he walked us to his truck. 'I'd been working on the telegraph poles at Marree. Forty-four degrees in the shade, got burnt too many times on all that metal. When they asked for a man to mend the phone up here I jumped at being near water. It was a week's job, but one look at Veronica and I never left. She just wrapped those legs around me. They don't make women like that anymore.'

The Roper looked too genteel for this outback wilderness. I wouldn't have been surprised to see a primary-coloured barge with geraniums turning the corner. Glyn leapt on to the jetty. 'This is my latest toy.' He switched on the engine of a flat-bottomed boat and showed us his sonar machine that pinpointed crocodiles and forty-pound barramundi swimming under the boat. A tail slithered off the muddy bank and fruit bats started screeching. Perhaps I wouldn't swim yet.

'It's half an hour down to the camps,' said Glyn. 'Hope you've brought some jeans, there are snakes out there.' We smiled weakly in our shorts. Within ten minutes, Glyn had carelessly pointed out a bank of red bee-eaters, a fish eagle, kingfisher and cormorant. 'But don't tell the tourists.' As we rounded the bend we saw an unmistakable cast-iron ship's boiler sitting upright and alone in the centre of the river. 'That's the *Young Australian*; they must have been on a bender when they parked her.'

Todd mentions the disaster in his 1884 report on the overland line. When he had evacuated in 1872, he'd left a group of mechanics and stockmen on the Roper to wait with the *Young Australian* for the iron poles. Within weeks they'd crashed her. 'You have to respect the river,' said Glyn, depositing us just upstream from the sunbathing crocs.

Picking our way gingerly up the bank through the undergrowth, I was looking for snakes when I saw something glinting. The seven of clubs. Frantic scrabbling produced another piece of tin, punched with the three of diamonds. We moved on to a gnarled, blazed tree disfigured with iron nails. 'The blacksmith's tree', said Glyn. Fragments of insulators, bottles and wagon axle hubs were buried under a thin layer of sand. It felt like a Roman encampment. Three hundred people once lived in Mason Town, now there was just metal among the leaves.

The Noah's knoll that had saved Todd's stranded men was half a mile from the Roper stores. An army field-oven with its door open was propped against a tree, left by a later wave of damp refugees. Telegraph Hill still operates as a refuge in the Wet.

We asked Glyn how often the Roper climbed this high. 'When the floods come, it's a lake from here to the Stuart Highway. That's how I pay for the boat. I'm a local taxi service for Ngukurr, the Aboriginal camp up the road. Deliver food to the stranded, take the ill to hospital, then I bunk off fishing the rest of the year.'

That night we took some barramundi steaks to our camper-van in a field. Two other tents were pitched next to us. In one was Lars, a multi-millionaire Norwegian who had taken six months out to fish. In the other was a Cypriot taxi driver from Brisbane and his mother. They let us share their barbecue and the Norwegian poured us a glass of claret. 'Black nights, silver barramundi, red earth and blue river, what more could I wish for,' Lars eulogised before moving on to vodka.

Ed and his shrimps set off with our new friends in their rowing boat the next morning, while I splashed across the Roper Bar, once the gateway to the Top End gold fields. Others after Todd had tried to colonise this paradise. By the early 1880s, a shanty town had grown up, filled with every insalubrious character in the district – men fleeing from the Territory's settled districts along the telegraph line and those who had been cast out of Queensland, 'the scum of society' as the Governor called them. In 1885, the government was forced to establish a police camp to keep the shootings under control. A hotel was built and a cemetery cordoned off. The place was named Urapunga. Only the missionaries dared come. By 1940, when the army set up camp, Urapunga was long deserted. Then in the early

1960s, Dieter and Joan Januschka appeared with baby Veronica. Now she had snared her man and the next generation was assured.

Ed returned with peeling arms and two small live barramundi in a bucket which he insisted on tipping into the coolbox so they could take pride of place. 'We don't need these Kraft slices anymore,' he said, hurling them to the taxi-driver's dog.

'But what about our sandwiches?'

I starved all the way to Katherine, the only metropolis between Alice and Darwin, and at least four hours away, but Ed was happily dreaming about our retirement with boat and camper-van on the Roper.

'We'd like a room with an oven, please,' we asked the motel receptionist in Katherine.

'We have a nice bedsit for six, room 143, but you can't take the fish,' she said.

'What do you mean?'

'You're hiding fish in there, I can smell them and they're not allowed in the rooms,' she said.

What about tinned tuna fish?

'That's different. No live pets is the rule.'

We thought of letting them escape into the swimming pool, but the receptionist was eyeing our cool-box.

So we drove them up to Katherine Gorge, where Todd's men used to bathe before the floods and holiday couples now watch the swans, to release them under the waterfall. 'You can't chuck those fish in here,' said the ranger. 'They'll die.' They were looking peaky anyway, so Ed bashed them on the head with a rock, and borrowed the ranger's knife to fillet them as a starter for supper.

The temperature gauge in the visitors' centre registered fifty degrees. 'Let's go for a walk,' I suggested. Ed looked doubtful. Minutes later we were clambering on the hot rocks, clutching hired lilos so we could float down through the gorge. After so many dry miles, we were obsessed by water. We were exhausted by the time we had clambered to the top of the first gorge, but the cool, shady float back to the camper was worth it. The next day, we went to Edith Gorge, double the length of an Olympic swimming pool and twice as clear, with a waterfall for the ambitious at the deep end. As we swam towards the falls, we could see fish swimming between our feet. We couldn't understand why the telegraph men had moaned so much about their conditions, but then this was the dry season.

'What about some basket-weaving?' I suggested when Ed started complaining that the Top End seemed too tame. 'This Dreamtime Brochure says that as well as learning how to throw spears and track kangaroo, we can try our hand at painting, and it's only an hour from Katherine.'

For $63 a day, the Aborigines at the Manyallaluk community teach tourists to walk in the bush, eat tucker and mix paints. We arrived at nine a.m. for biscuits and squash on rolling green lawns and met the rest of our group. 'Hi, we're Pino and Pina from Calabria, Adelaide,' said an Italian couple as we set off on our stroll. 'Keep up close,' said our Aboriginal guide. 'This is what the Aborigines use as plates, slates and loo roll, it's the most versatile plant . . .'

The guide generously said it wasn't just the telegraph that had ruined the land; the white man's greed for minerals hadn't helped either. His grandfather had worked in the local mines, his mother's family had been displaced from Arnhem Land. Many in his reserve had forgotten their backgrounds between the play and record buttons on their videos. Our guide wanted everyone to remember, and to prove that the Aborigines didn't need handouts. The camp was run as efficiently as a Club Med resort, and had won every tourist award.

After damper and kangaroo, we settled down to make the baskets with our guide's wife and sister. Some New Age travellers camping by the pond came to watch as we drew blood trying to get the reeds to split. Only Pina could master it. 'Pina's a natural homemaker. She makes fresh pasta for me every evening at our camp-site and she makes the best risotto with white truffles in the world. Only there are no white truffles in Australia,' Pino explained. They'd first come from Italy in the 1950s to raise enough money for a tractor for their village. Pino had joined the police so that one day the family could return. 'But when we went back to our village, they were richer than us. They all had tractors and harvesters. So we came back to Australia,' they beamed. 'We've been a thousand miles in our camper-van this year.'

Ed drew a fish in finger-painting class, I drew a tortoise. But it wasn't until the evening, when the others had left and we were rolling out our sleeping bags, that we relaxed. Two teenage Aborigines came down to share our fire with the New Age travellers. Did we want to listen to some stories? they asked politely. It wouldn't cost anything. Over the next few hours, we heard snatches of the dreamtime, as kangaroos became men, and men turned into lakes. Britain didn't interest them, nor did the travellers' anecdotes about life on the beach, so Ed recited Kipling's *Just So* stories about how the elephant got its trunk and the leopard got its spots. We had no wine, no cigarettes, just a few stale cigars and our language was miles apart. The fire had gone out, there were only voices in the dark, but in Kipling's rhythmical, ritualised language we forged our first connection with the native Australians and found ourselves invited back to their community to watch a video of *Four Weddings and a Funeral* the next night.

23

·· —— —— ··· —— ——

The Continent Connected

Todd hadn't yet experienced the dry season. By the end of March 1872, the Wet was almost over, but the humidity turned everyone's minds. Bushes and grass steamed as the sun broke through the clouds and sucked the water up. Clothes were just as drenched as they had been in the Wet, and infected mosquito bites were soon joined by huge, suppurating sun-blisters. Toads took over the camp in their thousands. As soon as the landing stage was clear, the *Bengal* headed downstream, taking the long-suffering passengers for Darwin with her.

Had the other camps survived, teetering precariously on their knolls without a Noah's Ark moored beside? Todd sent Patterson out to survey the land. It took him three weeks to travel the 120 miles to the line. Discovering Rutt's men first, he found them emaciated and down to their last biscuits. They had stopped themselves going mad by making a pack of cards from bully-beef tins, presumably like the ones we had found on the Roper. The men laboriously punched in the suits and numbers with a nail, and played euchre, gambling away their future salaries. McLachlan's camp was only a couple of miles away, but they hadn't spotted each other.

With Patterson's stores approaching from one side and Milner's sheep from the other, the men must have thought they'd seen a mirage and, after their first decent barbecue in six months, Patterson was able to get them working again by the middle of April. The total number of men employed on the northern section soon grew to about three hundred, including teamsters. They had over 250 horses, 350 working bullocks and 60 wagons carting supplies and poles. By this time, Little had managed to get the line between Katherine and Darwin working. Other men arrived at the Roper, having travelled up the line from the south, with optimistic news. So the project was running again, and they had six months before the next downpour. Todd and the remaining passengers left for Darwin in a festive mood. There was food and drink to spare, and everyone was toasted with champagne, including one for 'Mrs Todd and all the little Toddlings'.

But the bad luck soon returned. Patterson was making his way to Daly Waters when he met a man carrying dispatches to Todd. Patterson took it upon himself to open the letters and send his own assessment back. It was characteristically glum and pessimistic. Patterson insisted a pony line carrying messages across the gap couldn't start operating until at least August, and even then it would jeopardise everything by taking his best beasts. He wrote saying; 'The line cannot by ANY POSSIBILITY be completed by the end of September.' When Queensland heard the report, they immediately resumed discussions about their own line.

Todd was stuck on the *Young Australian*, making very slow progress round the coast towards Darwin. He had reverted to his original plan of riding down the entire length of the line. When the Postmaster-General finally reached Darwin he was mesmerised by its exotic smells. Chinese, Europeans and Australians all lived next to each other in squat wooden houses. The men from the British-Australian Telegraph Company (BAT), there to man the repeater station, had become the new elite. The company had realised that no one decent would stick the conditions without a few enticements, so had provided not only proper beds, a piano and curtains for the palatial new offices, but a billiard table, a well-stocked library and a pleasure boat for days off.

The Resident, Captain Douglas, presided over the entire affair with increasing amusement. He had slowly grown to enjoy his appointment. The Residency was finally being built and his younger daughter married the local BAT superintendent in the first society wedding in the Northern Territory. Todd wrote to Alice: 'He regrets having come to Port Darwin but admits it has brought him a good wife which is a great deal!'

Todd was less happy when Captain Douglas admitted he had no suitable horses he could sell him. Todd could inspect the line outside the back gardens, but he would have to go back to the Roper for transport. Even more irritating, he would once again be out of contact with the line, and so with Patterson. He had no idea whether his rival was following his orders.

Todd returned by sea to the Roper, where he met Patterson, who was livid that Todd was still interfering, and wrote in his diary: 'I handed Mr Todd four letters protesting against his actions after my departure . . . and informing him that if such action were repeated, I should at once resign the command of the expedition.' Todd again had to persuade him to stay, saying, 'I wish at once to say that I had no desire to fetter or embarrass you, or to interfere or over-ride any instructions you might have deemed necessary to give to the officers under your control.' This didn't mollify either man.

For an expedition to succeed, almost all explorers need luck. If Stanley and Livingstone hadn't bumped into each other in the Congo, one might have died and the other remained in obscurity. Everything had been against Todd and Patterson. Now the dice rolled their way. Just as both men were contemplating how on earth a pony express was going to be made to work, two brothers rode up to the southern end of the line with four stockmen and forty horses. The brothers had left Adelaide in January, determined to set up a stud in the new territories. The men at Barrow Creek, the repeater station above Alice Springs, couldn't believe it when the troop of healthy, shining ponies meandered into the backyard. They messaged down to Adelaide. The government sent back a jubilant note saying that they would lease a dozen.

Knuckey, who had been charged with organising the pony express, was coming down to find the end of the southern section with dispatches from Todd. He reached Barrow Creek, and transmitted the messages to Adelaide. That evening a dinner was being held in the town hall. As the meal ended, the mayor stood up and announced three cheers for the telegraph men, as the first trans-Australian message had just been sent by Todd. The guests raised toast after toast, little realising that the message had been carried for some of that distance by a pony.

The main gap was just above Tennant Creek in the northern section, and there were a few others higher up the line. In several places, crooked poles were the only ones available, and elsewhere wood was so sparse they had to make do with only ten poles to the mile. But the mood had changed; the men were now convinced that they could finish before the end of the dry season.

Todd decided that the time had come for the pony express. He had now travelled down to Daly Waters, from where he sent an urgent message up to Darwin for transmission to London, saying that the line was up and running. But because London had assumed the line was already working, several international telegrams were already waiting for the pony men, including one saying that Samuel Morse had died in New York on 2 April 1872. Todd was devastated that 'the father of telegraphy' hadn't lived to see the completion of the line. Todd's assistant, Cunningham, had also jumped the gun. On 24 June, an advertisement appeared in the Adelaide newspapers:

> Messages will be received at any South Australian telegraph office for transmission to London and other places in connection with the British-Australian Telegraph Company's cables during the ordinary office hours of Tuesday next. These messages will be forwarded from Tennant Creek by horse express over that portion of line at present incomplete, and are expected to reach their destination in eight or ten days.

The advertisement was written by a bureaucrat, not an ad-man, but the stilted prose couldn't conceal the drama of the moment. The telegraphists in Adelaide worked all night and thirty-eight messages were dispatched up to Darwin by the next morning. The official ones were free, but the private ones amounted to £423 11s. 9d. Adelaide would only receive £40 of that, with BAT getting the largest proportion. It was going to be a long time before South Australia recouped its investment. So far, the line had cost four times Todd's original estimate, and the bill now spent stood at £479,174 18s. 3d., not including the £3,000 a month that the government had been paying in fines to the company since January

Todd was baffled. The grand achievement of his dream may have been expensive and delayed, but its completion seemed such an anti-climax. After fifteen years of work, there wasn't even a drink for the men or a message of congratulation for himself. The men didn't know whether to raise a cheer, and Patterson insisted that many of the poles were already being eaten by white ants, and so the line might soon be one long pony express.

Only a few hours after the telegrams had come through from Darwin, Little was on the line again with a pressing message for Todd in Daly Waters. The unthinkable had happened. The under-sea cable from Darwin to Java had snapped and they had lost communication with Batavia. Todd's dream was broken. But this time it wasn't his problem. The BAT Company could soon be paying Adelaide if it couldn't locate the fault. A couple of days later, news came that the break was somewhere in the ocean, and a ship had been sent out to find it.

Todd, concerned that they would soon be sitting on a stack of useless messages, sent one of his telegraphists, Ray Boucaut, down from Daly Waters to Tennant Creek, to warn Adelaide to halt the traffic. Boucaut, more used to an office than a saddle, rode flat out to preserve the Post Office's reputation. He covered 262 miles as he rushed down the route, and was in the saddle for 100 hours continuously before he reached the completed southern end of the line. There he sent down the first telegrams from England and passed on the news about the break in the line.

The colony was despondent, but not for long. On the same day, a letter arrived saying that Lord Monck, Chairman of BAT, was so appalled by South Australia's delay that he had given Queensland the go-ahead to start laying a cable to Norman River from Port Darwin. For once Lord Monck's BAT Company had been wrong-footed. With their line broken, they could no longer play the two colonies off, but would have to grovel to South Australia. The company, said Todd, was now 'hoist with its own petard'.

The ponymen already had one sheaf of messages from Adelaide, and rode across the gap to meet Todd and Patterson who were making their way down towards Tennant Creek with the surgeon Dr Renner. Todd asked the ponymen to stay up at the northern end until they could bring the message down that the cable had been repaired. The superintendent's mood lifted as he travelled down the line. At every camp and station, he received letters of congratulation signed by the men.

The poles, he admitted, would need to be slowly replaced by metal Oppenheimers, but this work had already started. Other problems were almost untangled and the line would soon be complete. Todd started punning again, always a good, if embarrassing sign. Pointing to one of his companions, Wells, he would tell the working parties, 'I cannot be affected by the next stretch as I carry Wells with me.'

Patterson was less relaxed. 'Mr Todd is getting awfully tiresome and I am looking forward with unaffected delight to seeing the last of him. He should have great reason to be downcast about the line . . . I would not be in his shoes for double his salary.'

By mid-July 1872, Todd arrived at Attack Creek, just above Tennant Creek, and found the wire was now operating all the way up from Adelaide. He attached his pocket relay to the line and started sending messages back to his office. Of equal relief was Patterson's agreement that he would leave Todd to return to the northern working parties, and Todd would continue to make his way down south. Patterson had been feverish for the past two months. His eyesight was increasingly poor and he complained of headaches. Todd was unsympathetic, convinced he was faking it. In fact, Patterson was probably working himself into a frenzy at the thought that Todd might be the one to have the glory of joining the line.

When Todd went south, Patterson was rejuvenated, and wrote: 'Twelve months since I left home. All difficulties, and there were many, are overcome, and the battle is won.' But his black humour soon returned.

In Adelaide, they were equally triumphant. Lord Monck, still unaware that his cable had broken, had just sent another message, insisting that BAT had every right to find an alternative route now that it was apparent that the overland experiment had failed so dismally. Lord Monck wrote: 'The fact that a capital of £600,000 has been lying without any return to the proprietors, or use to the public, for a period of seven months, is sufficient to justify the Company in making every effort to put a stop to so disastrous a state of affairs.'

The politicians and newspapers couldn't contain their glee. An editorial in the *Advertiser* ran: 'We believe that the grand secret of the scurvy way in which we have been treated is the fact that the Construction Company must go on manufacturing cables, and to find a demand for these they must promote new telegraphic enterprise. The more they can do in this way, the more grist they will bring to their own mill.'

Even better, on 9 August the last pole was planted, and all that was needed was a few more coils of wire. Patterson panicked that he was now too weak to reach the joining point. Todd, now at Barrow Creek, contemplated making the link himself. He sent a message to the government saying that there were now only two short gaps in the wire, and asked whether they would like a formal opening ceremony. The politicians decided that any official junketing should take place in Adelaide, presumably so they could enjoy the acclamation of the public without a testing trip inland.

There had been sixteen changes of government, with twelve different men as Premier since Todd had first floated his plans in 1857. Milnes, Hart and Blyth had missed their chance, and now the government of Henry Ayers wanted to ensure it reaped the benefits of its predecessors' investment of money and reputation. Todd couldn't begrudge Ayers, one of the few people to believe in him, a share in the spoils.

Patterson, now in sole command, insisted the men cut the new line near him at Frews Pond, so he would have little problem in getting out of his tent and linking the two last pieces. He sent word to Adelaide that he would make the link at midday on Thursday 22 August. Todd told the Post Office to raise the red ensign on the roof as soon as the link was made. The men had to wire in the dark to meet the deadline. On the Thursday morning, Patterson struggled on to his horse and, joined by ponymen and several linesmen and a telegraphist – nine in all – made his way to the cut.

The telegraphist used his pocket relay to ensure that both lines were prepared. Patterson chattered away nervously. At two minutes to noon the two sides were ready. Patterson seized one end of the wire, and half a dozen men seized the other. Heaving and straining they tried to bring the two lines together. It was forty-two degrees, sweat was making their hands slippery, and Patterson was now jabbering to himself maniacally. The enforced cut had shrunk the wires, and the ends would no longer meet. The men began to laugh at the comedy.

Patterson, in a fury, dropped his end and ran to get a length of binding wire. He wrapped it around one hand, and then made the connection with the wire. The telegraphist frowned. Just as he shouted a warning, a piercing scream rent the air. Patterson was flung to the ground. The electric current from the batteries at the stations had flowed down the line and through Patterson.

At first there was no movement, then his fingers began to twitch. Patterson gingerly rolled over and, brushing aside the men's attempts to help, stumbled up. 'Next time I proceeded more cautiously,' he says in his diary, 'and used my handkerchief to seize the wire. In about five minutes I had the join made complete, and Adelaide was in communication with Port Darwin. It would have been with England had the cable not broken down.

We then drank success to the Overland Telegraph, and fired off 21 rounds from our revolvers and immediately adjourned to Frew's Pond.'

The men broke a bottle of brandy over one of the poles. Patterson returned to his bed. The message went down the line and soon all the teams were opening the rum bottles. The under-sea cable was still down, but that wasn't their problem. They were exhausted, elated and free.

Yet they couldn't draw themselves away easily. For just under two years, from August 1870 to July 1872, they had been living the frontier dream – a baking paradise or a muddy hell – and only five men had died. Some wanted to return to their families and businesses, but several couldn't bear the thought of their mundane clapboards in the city, and decided to stay and try their luck as gold prospectors. Others pegged out grazing land to lease, and a few remained to man the repeater stations or to put up the new poles. Patterson laid on a steamer at the Roper to take anyone home for free before the wet season recommenced.

The continent was now connected by a thin metal wire you could almost floss your teeth on.

The men who stayed opened up Australia, slowly building a civilisation in the outback. Their towns were the repeater stations, the line their supply train. In a few years, they transformed the heart of the continent. The telegraph line produced a rash of exploration, as Europeans began to criss-cross the arid lands. Within fifty years there were ice-cream parlours in Alice. The lives of the original inhabitants, the Aborigines, had been irreversibly changed. Their waterholes were expropriated and they were hounded by law enforcement officers. The line had not only linked Australia with the old world, but had opened up a new world for pastoralists, missionaries and gold diggers.

In Adelaide, the Todd children watched as the red ensign was raised on the Post Office tower. The Town Hall bells rang out, flags were hoisted in the suburbs, the ships in Port Adelaide flew bunting. Pupils were allowed out of school, office blocks emptied, stores closed. The day was declared a holiday.

And where was Todd? At Central Mount Stuart, the geographical heart of Australia. The Superintendent of Telegraphs had decided to camp at the foot of his favourite spot, to savour the moment. As the sun set, he nervously clipped his pocket relay on to the line.

The whole of Adelaide responded. The Chief Secretary was first to reply. 'We opened the line at one p.m. this day as it was completed,' he began, and went on to praise the men, 'for the praiseworthy efforts and untiring diligence that they have displayed in bringing to a successful conclusion this great work, under your able superintendence. Accept my congratulations that your troubles are now over.'

Todd replied: 'Many thanks for your kind congratulations on the completion of the telegraph, which, as an important link in the electric chain of communication connecting the Australian Colony with the mother country and the whole of the civilised and commercial world, will, I trust, redound to the credit of South Australia.'

Patterson was forgotten. He didn't know how to converse in Morse code. Todd, with his tiny line-relay, sounder and Morse-code key, sat reeling in the messages from politicians all over Australia, friends, and colleagues, and, finally, his nervous wife, who had been taken to the post office by her children to send congratulations to their father. Todd chattered away until midnight. 'I transmitted my replies until I was nearly frozen and completely knocked up with fatigue and excitement,' he wrote in his diary. 'Thus the great work, notwithstanding all our mishaps and disasters, was successfully accomplished within two years.' Then he packed up his portable, rolled out his blanket and fell asleep.

24

·· — — —   ···· —

Adelaide-in-Waiting

The next morning, Alice put on her lilac satin dress and followed the crowds the half mile to the post office to watch the first messages coming through. On the notice board were the daily weather reports from the colony and right at the top was Darwin – fine, cool and clear. One year and eleven months after the first men had left, the line was finished.

For almost all that time, Alice had been wearing black. Still only thirty-six, she kept returning to the image of Stuart, his hair turned white, hobbling down the street after his epic trek. Would she recognise her husband, and would he recognise her? Her mother tried to insist she stay out of the sun, but the garden was her only retreat from the snubs on the street. Alice knew her face was now sprinkled with brown freckles, her wrists and ankles were bloated from too many comforting puddings and her forehead was etched with lines. Her Cambridge cousins wouldn't recognise her.

As she slipped into the new post office that had just been finished on King William Street, a few began to clap. Then more joined in. Soon everyone was cheering. Overwhelmed, Alice fainted on the polished wooden floor and had to be carried up to Todd's office. She was rescued by her eldest daughter, Lizzie, and taken home to bed. For three days, Alice was delirious with a fever, worrying that Todd would have to pay the difference between his estimated cost of £120,000 and the £400,000 now being bandied about.

Alice had written regularly to 'My own dear Charles' on the Roper, and he had done his best to reply. But her first few letters show her 'in such low spirits', according to Lorna, that, when she was going through her mother's correspondence, she felt it was her duty to burn them.

Alice herself writes in one missive: 'Take care to burn my letters, I don't want them flying all over the earth.' She kept all Todd's letters in an old embroidery bag. Pat said Lorna would never discuss the contents of her bonfire of letters, but she thought Alice was suffering from the depression that had dogged her since Maude was born. To make matters worse, old Mrs Bell was constantly reminding her daughter Alice of the sacrifices she'd made to be with her in Adelaide, and scolding her for letting the children run wild. Mrs Bell could have stayed with her son, Edward, at his new house in Chesterton, where tea was served promptly, they knew how to respect their elders and her hair didn't go frizzy in the damp. Lorna wrote to one cousin: 'To make matters worse for my mother, my grandmother came out to stay. To say that she disapproved of everything is to put it mildly.'

Other wives, with their husbands away on the line, would try to jolly her along. But it only made Alice feel guilty. What if the project did fail, or the teams were hit by disease? She was determined to be stoical in public. But when word came from a passing ship that Todd had caught a fever, Alice worried that she might be widowed by her husband's dream.

The first letter I found in the Mortlock Wing of the State Library, tied up by the librarians in pink binding tape, was one from Todd to his sons, Charlie and Hedley, urging them to work hard while he was away on the Roper.

I am glad that you and Hedley are about to learn drawing . . . Remember, my dear boys, now is the time for improvement. Many boys no older than you have to go out into the world and earn their living, but you and Hedley will, I hope – should my life be spared – have several more years at school. I know you will both persevere, and by diligence and assiduity do all you can to secure a first class education such as I wish to give you.

The anxious Victorian father then relaxes a little. 'The river abounds in alligators of large size and sharks, we caught a number of them yesterday – also cat fish which we cooked but I can't say I liked them. They were like the Bishop's apron, they went against the stomach . . . Your loving Papa, Charles Todd.'

The boys, who were obsessed with horses and betting, were already running circles around their teachers. They had nicknamed the baby, my great-grandmother, Nina. The children carted her around slung over their backs in a blanket. Maude would still only wear her brothers' trousers and seventeen-year-old Lizzie was desperate that she would never fall in love. Ever since her father started the line, her social life had dried up and her mother's headaches were now continuous. She wrote to her father about attending Mrs Bird's Bible class, the death of her cousin Fanny's baby and the celebrations for the Queen's birthday. But she couldn't hide her hurt at the family's constant rebuffs. 'The Wearings have taken a house in North Adelaide. Mama went to call upon her today but she was too busy to be seen,' Lizzie writes. 'I think it was very rude of Mrs Hart not to invite Mama to their ball tonight, but we all think that the Roper has not agreed with Mr Hart and that is the reason she was not invited. Mama does not think it is Mrs Hart's fault, as she has always been so very pleasant to her,' she adds.

By the time the floods were subsiding on the Roper, Lizzie had made two new friends through the church and so found a way to socialise. On 29 May, she wrote, 'Mama and I thought it would be very nice if I could go out sometimes with Dr and Mrs Miller. As they go out so often I might then have a chance of getting out more. Last night after bible class I went in to see Mrs Symes and stayed to tea. She is so jolly.'

Alice no longer cared to go visiting where she wasn't wanted, and she couldn't bear the frivolous way in which the telegraph line was now being dismissed as a white elephant. Instead, she sent long, rambling letters to Todd. The first surviving letter was written when Todd had already left the Roper. 'I have not heard from you for more than a month – five weeks tomorrow and it seems such a long time. I hope by this time you have commenced your journey home at last, and may God in his kind mercy carry you to safety to your home . . . the time has been so long.'

After her initial despondency, Alice decided that if there weren't going to be visitors, she would throw herself into good works. Her first target was the wife of Alfred Symonds, Patterson's storekeeper, who had been convicted of felony. 'The fact is, poor thing, she married against the consent of her family and they do not wish to recognise her now. She has five children . . . I went to see the poor creature, she has only two chairs and a table in the room and she receives rations from the Government . . . The world is full of distress,' she tells her husband.

The church and the new vicar, Mr Symes, soon become a lifeline for Alice and Lizzie. While the other children would slip off to the races, they would go to services three times a day on Sundays. 'They have been erecting new meeting rooms and classrooms,' Alice writes. 'Mr Hay gave £100 . . . I said I thought you would give £10, I did not feel we could possibly afford more. I hope you will not be vexed. Charlie was with me and when Mr Hay said he would give £100, Charlie said, "what a nice horse that would buy." I like the Symes so much, they are people of refinement and good taste.' She adds,

We are all quite well, Nina has a little native pock about her neck, only a little fractious. She cannot understand why her father does not come home and kiss her . . . I hope you will be pleased with them all. It is a great anxiety for me to arrange without you my dear one . . . We are going, i.e. Lizzie and I go, with the Millers tomorrow evening to an entertainment in aid of the Building Fund of the Stow Church. The first time I have been out of an evening since you left. How very thankful I shall be to hear a message from you.'

Slowly, as word leaked back that the Wet had been conquered and the line might one day be completed, carriages began to call at the observatory. 'Today I am going to call upon Miss Davenport,' Alice writes. 'I have called on Mrs Barrow, she was so pleasant and dear . . . I have seen Mr Ross back in town, he always stops to speak.' By 12 June she was quite jaunty. The gossip was that Todd had pulled it off. 'Everyone seems to be getting very excited,' she writes. 'Mr Cunningham went to see the Chief Secretary and he thought there should be some signal given to the public when the first message arrived, and Mr Ayers at once ordered a splendid large ensign to be made for the tower of the post office.'

Then the invitations started arriving. 'The Hughes of Torrens Park have invited me to a large party on the 21st but I have declined the invitation. I should not like to enter such a scene of gaiety whilst you perhaps are enduring such hardships.' After the first trans-continental telegram arrived in Adelaide via the pony express, Alice's calling cards soon became cherished commodities. The children watched as carriages queued up outside the gates. Everyone wanted to know if she had heard any news from her husband and whether there would be cut prices on the line for friends. Alice, still nervous about being rejected, spent most of her time with her new church friends. She writes, 'I have been down to see Mrs Miller . . . she is such a dear young thing, so cheerful and affectionate. She knows me so well and laughs at me and talks in her pretty, crooning way.'

Money was still a problem. Todd's junior, Mr Cunningham, was looking after their financial affairs. Alice, who had always adored shopping, hadn't bought a frippery for a year. Charles wrote, 'Dear Alice, I know you will be economical, I want to find a good balance at the bank when I return.' But Alice was already showing more financial acumen than her husband. 'The Bank of Adelaide has issued New Scrip . . . I have made a deal with Mr Cunningham about buying some as I believe you would not let such a good chance slip,' she writes.

She was also concerned about the properties that Todd had bought only the year before. Just before the line was started, he had gone into partnership with her niece Fanny's father-in-law, Dr Charles Davies. Together they acquired two properties called Mattawilungula and Moonarcoe in the Gawler ranges, west of Port Augusta. Todd said he had no illusions about playing the squire, but he'd like some property for his sons so they could feel proper Australians. When he went to inspect his new land he found it covered in spring wildflowers, but soon the property manager was sending back reports complaining that there wasn't a blade of grass left, just mallee scrub. Alice realised she didn't even know where the properties were on a map.

She found she quite enjoyed managing her husband's business affairs. 'I went to the office to see Mr Cunningham about the shares a few day since,' she writes, 'and I saw your nice rooms. They are keeping them wrapped up during your absence. The carpet has cloth over it to prevent dust going through, and camphor is strewn all over it so the moths may not touch it.'

A week later she got her first invitation for three years to spend a few days in the country at Mount Barker. Lizzie had also been invited to her first ball, so needed a new dress. The boys had been reading *Bleak House* at their Bible classes and 'there is now nothing but talk of England', according to Lizzie. Their mother was prepared to sit for hours discussing her homeland.

As Charles took months to wind his way down the line, however, Alice slipped back into depression. One day she would mount a map of the overland telegraph in the hall so she could monitor his progress, the next she would retire to bed and start cutting up her Limerick lace wedding veil as curtains for the dolls' house.

Then, at the beginning, of July, Alice thought she had seen a sign from God. 'Oh how I have longed for you lately my darling, I want you so often more than ever. You will I know rejoice with me that during the last few days my mind which has been distressed for some time has at last found grace in my saviour,' she writes to Todd. 'I went to Mr Symes and had some chat with him and he showed me that I myself had nothing to do but believe God's holy word which says, "him that cometh unto me I will in no wise cast out." I trusted that God could not be worse than his word, and I have had such peace.'

After her Damascene experience, Alice writes that her mind was continually being fed with spiritual things. She believed that God would help her come to terms with living in Australia, and throw off her melancholia. 'But the last few days I have again been disturbed. I am so fearful lest Satan should again gain hold of me, but I trust and I also pray that the Divine Arm may be round about me . . . You cannot think what a comfort it would have been if I had you to talk with, you would have helped me so much.'

Alice was soon going to services every day, and the family visited the destitute asylum every Wednesday and the city mission every Thursday. Partying was now frowned upon. Alice was inspired by the nuns and there were plenty of sad cases on which to practise her charity. 'You will, my darling, be so sorry to hear of the sad accident which has happened to five of our fellow colonists, including two of Mr White's sons,' Alice writes to Todd. 'They all went out last Tuesday fortnight on the bay. The boatmen told them that the night would be a dirty one . . . but they would persist in going . . . and ever since the boats have been looking out for some clue. Mrs White has been prematurely confined and it is said that the dear little baby does nothing but moan.' Lizzie was soon dispatched with a basket of jams and cakes.

One piece of gossip that was consuming her new Bible friends was the dalliance between the exemplary Miss Heaton and the old Reverend Austin. Miss Heaton was Alice's role model. She is described as 'extremely clever, makes good use of every little bit of material and uses them for the poor. She gives away so much of her income.' Alice is less sure about the Reverend Austin. 'He is 72 and not at all well off. But I fancy she will have him,' she tells her husband.

Lizzie was turning into the catch of the season. 'Maggie Milne is to be married next Tuesday and there is to be a large party,' Alice writes. 'Liz is invited to the party and I wish she were not. I am going, I do not care for it but I do not like my child going without me.' Alice perseveres: 'I have refused the invitation to the Bachelors' Ball. I do not care to go without you. I should now feel myself strangely out of place.'

Alice was still flustered about the money. Her mother's nest-egg had run out, and it was only now that she had any inkling of the state of their finances. 'There were a great many bills left for me to pay and I have done the best I could,' she writes. But the family hadn't bought any new cloth for a year. Grandmother Bell had made the boys new school uniforms out of two pairs of Todd's old trousers, but there were the tutor, the maid and the pew rent still to be paid.

By the next month, Alice, once so keen for social acceptance, had decided that dancing was devil's work. 'I do not care for large parties at all. It seems to me such a waste of time and it seems so unsatisfactory a way of spending money.'

Alice was determined to share her new-found religious fervour with her husband. 'I'm so sorry, now that I have found the delight of reading religious works, that you did not take more nice reading. I do so enjoy reading God's word.'

The remaining letters, written in a childish, round hand, suggest she never seems to have regretted marrying Todd. 'When I really have you with me I shall spoil you my own darling. I do indeed long to give you one fond kiss, and I will have more than one,' Alice tells him.

On the day that the underwater cable went dead she became re-confirmed into her new church with a special ceremony. 'I wish my own darling you could be with me on the evening I join the church and the first Sunday I partake of the Lord's Supper . . . they are having special services at Mr Mead's church for the unconverted. I get the servants to go sometimes.'

On 29 August, when the flag had been hoisted and she had received her first telegram from her husband, Alice wrote to Todd to congratulate him upon his success. 'The congratulations by persons calling and by letter have been overwhelming,' she writes, omitting to say that she had fainted. 'The first who called was the [Governor's] private secretary early on Friday morning and he was so pleasant. Then later in the day numberless friends called. All were so hearty I felt quite overpowered.'

Suddenly, everyone wanted to know Mrs Todd, and there was barely time for prayer meetings. 'All say I shall not be able to see you for the first day,' she writes. 'The public are going to have you all to themselves, but I hope I shall get a sight of you by some means or other.' Neighbours soon overcame their embarrassment at their previous reticence and Alice was asked back into the smartest drawing rooms. 'I went to the floral bazaar on Tuesday. There I had a chat with Mrs Bruce,' Alice writes. 'She said to me "Where have you been all this season?" I told her I did not care to go without you. But she said, "You will not have that excuse long now that Mr Todd has completed the line." All say the same thing . . . Mrs Barr-Smith called a few days since and she said, "Well Mrs Todd, you ought indeed be proud to have a husband who has so distinguished himself . . . All Adelaide seems proud of you and so they ought."' Even Lady Charlotte called. 'She has been to see me two Sundays. I wish she would keep away on Sunday. I like to rest with my dear ones. We have now gone through Genesis and are reading the travels of the children of Moses.'

The good works were still squeezed in. 'I went to see Alfred Symond's wife a few days since and my heart ached to see the misery,' Alice writes.

She has two rooms, one of which is so dark that you could hardly see in broad daylight . . . when the window is open the smells from a closet close by and the remains of vegetables that are thrown out make the place most unwholesome. I went to see another poor creature who is in a decline and it was such a change, everything was beautifully clean and so was the poor woman, but underneath the room in which they slept was a cellar which is filled nearly to the top with water and you can see the damp rising . . . It makes my heart ache to see such misery but I am sure it is wrong to shut one's eyes and not find out the misery there is around one.

Alice had, however, dropped the idea of emulating Miss Heaton and giving their money away. She desperately wanted to go on a proper holiday – 'a late honeymoon' – in Tasmania or Melbourne, and she had scribbled down a list of improvements for the observatory. 'So many say to me you ought to have £5,000 from the Government and 12 months leave of absence,' she writes. 'If you get £2,000 it will be very nice. I should not expect more, should you? I would like to have two or three more rooms built as we are fearfully short of rooms. And I should like a dressing room.' Alice planned endless expeditions with her husband when he returned. She imagined cosy evenings, picnics at Glenelg and visits to the Adelaide hills. She also wanted to buy some more material so she could make her husband some new trousers with 'continuations' which were now all the rage.

Todd wrote back, 'I would like to take you and dear Lizzie to Melbourne and Tasmania and hope the Government will be sufficiently generous to justify our taking the trip together. I would like us to go to England but must not think of that just yet.' Concerned that Patterson would try to make mischief on his return to Adelaide, he also sent a dispatch to Alice asking her to see Elsey, Patterson's wife. Alice obviously liked her. She writes, 'Mrs Patterson had such a laugh with me about her husband's trowsers. She seems a nice little woman. She has been married three years, and eighteen months of that time they have been separated . . . She has a nice little cottage opposite College Avenue and the drawing room is very prettily furnished.' More appealing to Todd, she adds, 'The judge says you must be knighted and get a handsome bonus too.'

Once Todd had passed Alice Springs, Alice could regularly contact him on the new line. Todd tapped out that he had only a spring cart and a horse to make it home. Alice described what she wanted to cook him as a celebration meal. But they often found themselves tongue-tied with the new machine. Alice felt so embarrassed pouring out her affections to a post-office scribe that she returned to writing. 'Lizzie's taste in dress is neat and plain, no feathers or flowers, and no bright colours in the street. I think she is very near being a Christian.' Charles is a born lawyer, and Alice has suddenly noticed her youngest daughter, Nina, calling her 'a round little ball'. She pretends to be hurt that Todd has messaged Cunningham to ask him how she looks. 'How could you be so impudent?' she writes. 'Well, dearest one, I take tremendous walks but I do not get any thinner, but as I am so well now, I do not trouble myself.'

The government had been muttering about giving the title of Postmaster-General to a ministerial officer when it looked as though the line had failed. Todd was desperate to know whether he would now keep the title. Alice reassured him: 'You will see the kindly feeling the Government has for you and you will see that nothing would be done to annoy you. I believe, and all say so too, that whatever you want you will have.'

Eighteen years after Alice had proposed to her husband, her letters show that she was still very much in love with him. 'I hope we shall never be separated again, it is so hard,' she writes. 'My anxiety about you has sometimes been unbearable. When the first letter arrived telling me you had fever, I thought I should lose my reason – I don't know how I did get through the first few months of your absence . . . Nina says when you come home she intends sleeping with you, but I don't think I shall agree to that. I shall sometimes want to have you all to myself.'

When Charles had reached the Peake, she wrote: 'Do not disappoint me now. I am afraid if you are not home by the time you mentioned I shall be ill with excitement.'

Todd immediately reassured her, writing: 'I cannot by telegraph say all that I could wish, nor express myself so affectionately.' He hadn't received any of Alice's letters until he reached Charlotte Waters in August, once the line was completed.

Todd was most struck by Alice's new religious fervour, hoping perhaps it would alleviate her depressions. 'My heart is filled with joy and gratitude at the decided step you have undertaken which with God's blessing must exercise an important influence on our future life in this world as well as all our children whose characters have now to be formed. So much depends upon our example . . .Your letters are so loving they make me long for home.' He worried about her 'neuralgia', and admitted that his journey has been 'fatiguing and the absence of a nutritious diet has not tended to strengthen me'. The answer, he suggests, may indeed be a trip to Tasmania. He sends her a box of the Aboriginal coral necklaces he acquired on the way, flouting his own rules on trading with the Aborigines.

Another letter starts: 'When I get back I quite intend to take things more gaily, take days off, go racing, I should like to be able to rest on my laurels a bit.' Alice was thrilled. She even dared to think that, now the line was completed, perhaps they might go home. Her husband could get a job in Cambridge, or she would be happy to move to London. The children could go to British schools. Lizzie could come out in society.

She enclosed several sermons, and relayed the news that a university was soon to be built, 'so nice for our two boys' if they stayed. Her visits to the poor had not been going as well. 'I often wish when I visit poor people that I had not such a fussy nose. I always feel so queer and sick when there is any bad smell.'

Thirteen days before Todd arrived home, Alice admitted that she was longing for him to return so much that she was afraid to think about it. 'I was quite ill when you went away and when your first letter arrived,' she admits. 'I did not know what to do with myself, I was so fearful lest the climate should so weaken you that you would not rally . . .We shall now, my darling, be more thoroughly one than ever.' She even played a game of Pope Joan. 'I don't now care for cards, but I did not think I was doing wrong by playing.'

After ten months sleeping rough, Todd writes that he is now so attached to his hammock that he wants to suspend it between the GPO and a lamp-post. 'It won't hold two you know,' he jokes. The last letter in the library from Alice reads, 'I don't think I shall like a hammock. I am perfectly satisfied with my beautiful bed. I often think how pleasantly the lines have fallen to me. I have such a beautiful home and soon I shall have my own dear husband. Good night darling, with fondest love from your own loving wife, Alice.'

25

··— — — ·····

The Grand Electric Chain

As Todd rode the final miles home, the BAT cable company was frantically scrabbling along the sea floor, trying to find out where its line had failed. The smallest hole would be enough to let in seawater and rot the copper wires. The operators on the *Investigator* had to scan the entire route to discover where the break had occurred. They found the malfunctioning section near Banjoewanji, but the first two times they raised the line it snapped. After three attempts, they finally managed to splice in a new lead.

Two weeks before Todd reached Adelaide, he intercepted a message at Beltana. The dead cable had come to life. On 21 October 1872, international communication was restored after four months' interruption. The Great Work was finished. 'The Australian Colonies were connected with the grand electric chain which unites all the nations of the earth,' Todd wrote.

The first private message came in from London to Melbourne, and in the next batch was a congratulatory cable from the Lord Mayor of London to the mayor of Adelaide, sent only seven hours earlier across a distance of over 12,500 miles. The town raised the flags: Australia had joined the Victorian Internet.

A leader-writer wrote in Victoria's *Argus*:

Thus the long line we have been following across the wreck-strewn bottom of the Bay of Biscay, in the blue depths of the Mediterranean, down in the heated waters of the Red Sea, across the broad stretch of the Arabian Sea, through Central India, again plunging into the sea in the Bay of Bengal, treading the channels of the Straits of Malacca, crossing through the rich tropical scenery and amidst the towering volcanoes of Java, and thence once more diving down into the coral depths of the ocean, finally makes its landing on the low mangrove-covered shores of North Australia.

Todd left his horse at Burra to catch the train to Adelaide on 30 October. He asked for no fanfare, but the Chief Secretary went to meet him at the station. At first, few recognised this weather-beaten man as their telegraph chief, astronomer and Postmaster-General, but soon everyone at the station was crowding round. The *Advertiser* wrote, 'He has made his mark upon history which will never be obliterated.' The *Register* called for three cheers for 'the moving spirit in this gigantic undertaking'. The *Sydney Morning Herald* called the line, 'the last great triumph of telegraphy' and Victoria's *Argus* said the telegraph would link Australians 'more closely to the family of nations'. The trials of the past two years were momentarily forgotten.

When Todd rode in to inspect his newly completed post office for the first time, the men hung garlands of flowers above the windows and erected a floral arch at the entrance. Already, 153 messages had been sent from Australia and 148 messages received, yielding total receipts of £3,068. South Australia's share may only have been £353, but at least the line was beginning to pay its way. Within a year, South Australia had received £12,000 from cable messages, plus £3,600 from local traffic. The minimum cost of a telegram to London was £10, but many were over £100 as wheat growers and mining companies soon discovered the line's true worth.

Within hours, they could now ascertain the best markets. In only six months, arable farmers had already made an extra £250,000.

To achieve this, Todd's men had traversed over 2,000 miles, through a continent that had been crossed only once before. They had laid over 36,000 poles and transported over 2,000 tonnes of material, including batteries, receivers and stone for the stations. More than 3,000 sheep and cattle had for the first time been driven across the centre. When the government tried a similar feat with the railways ten years later, one man was stabbed to death in the first week, another was accused of rape in the second, police had to be sent up the line to control gangs of looters, and the railway still hasn't been completed a hundred years later.

The *Advertiser* decided to test out the line for its readers. The article began, 'We accept the fact that upwards of 2,000 miles of line is being carried through the heart of the continent . . . but how many of our readers grasp the fact that our communication with Darwin is practically instantaneous?' As an experiment, the paper sent a list of questions up to the Northern Territory. 'In several instances the questions containing many words were transmitted, and the reply obtained in less than 30 seconds! The mysterious agent passed over upwards of 2,000 miles of wire in less than half a minute . . . It seemed more like a miracle than a sober fact that we should be holding intercourse with friends at such a distance.'

The Australian Associated Press linked up with Reuters to provide a daily telegram of European news, supplemented by their own agent in London sending special messages. After a week of using the cabled news, the editor of the *Sydney Morning Herald* took stock. 'Of what use is it?' he asked. 'How much worse off would Australia be, if the line failed?' He decided that the telegraph was a great dissipater of anxiety. 'Around the whole civilised universe, the telegraph, like a sentinel, moves with the same announcement, and says, day by day, of some great interest, full of danger, anxiety, and importance to millions, "All's Well." '

Within weeks, most urban Australians had forgotten that they used to read news that was three months old. Editors, who were footing the bill, had to remind their readers just how lucky they were. 'Subscribers . . . could scarcely realise the fact that they were reading of occurrences that had taken place within 24 hours, at a distance of some 12,000 miles; but people of the present day soon lose their wonder of the marvellous,' the editor of the *Sydney Morning Herald* wrote.

Alice may have thought that, after eighteen years, she would finally come first in her husband's affections, but Todd barely had time to see his family. There was the new post office building to inspect, stacks of reports from his subordinate officers to read, the weather reports to be updated, and he had completely overlooked the Observatory for the past few years. The busy functionary never quite had time to squeeze in outings to the beach, and he looked distracted during the Reverend Symes's sermons.

There was still work to do on the telegraph. Most of the men, led by Patterson, were making their way home from the Roper by sea on the *Omeo*. Some had been away for two-and-a-half years. They laid into the *Omeo*'s supplies of beer and champagne. Captain Calder didn't mind, he needed the empty beer boxes for extra fuel to get home.

Arriving back, the men in their blue shirts and moleskins queued up at the GPO to draw their pay cheques in the biggest pay-day in Australia's history. As the town's shopkeepers and barmen rubbed their hands in anticipation, Adelaide planned a full day of celebrations in honour of the men. On 15 November they were told to meet Todd in the main hall of the post office. At eleven-fifteen a.m. the town bells started ringing. The men began marching along King William Street to the Exhibition Grounds. When they had gathered, Todd stood up to give a speech, thanking the men for helping him achieve the proudest moment of his life.

I come here to say nothing of myself; I have borne but a small part in this important work; but I come to bear testimony to the able manner in which our plans have been carried out by the brave men . . . They have performed deeds of heroism and endured patiently the hardships they have had to encounter; and no matter what privations they may have been suffering, they always gave me a hearty welcome . . . I am very much indebted to them for carrying out to a successful issue this great work in which my reputation has been at stake.

Patterson was too ill to attend, so Charles raised three cheers 'to the man who has done so much to promote the success of the expedition'. The band struck up 'See the Conquering Heroes Come', and the families cheered.

After lavish picnics, roasted pigs and kite-flying, there were sports competitions all afternoon. In the evening, the men had been given a ticket for a banquet at the town hall. Guests and dignitaries were allowed to attend for a guinea, and women were given permission to look down from the gallery.

In Sydney, Darwin and London, parties were held. Sydney's new Governor, Sir Hercules Robinson, made a long speech in which he barely mentioned South Australia. The *Sydney Morning Herald* pointed out that 56 of the 210 men who had just returned were actually from New South Wales. But the Colonial Secretary and Premier, Sir Henry Parkes, insisted that South Australia should get the most credit. 'The Government of which I am a member is not privileged to claim any credit. It has remained for one of the Colonies, by no means the largest in resources or population, to achieve this great and signal triumph,' he said. As another speaker rose to belittle the southern colony, Sir Henry intervened again. 'The thanks and sympathy of every one of these colonies and of every person in these Australian communities, are due to South Australia for her brilliant pluck and indomitable perseverance in piercing the central desert of this country.'

London was also holding a Grand Telegraph Banquet at the Royal Colonial Institute. Throughout their day and Australia's night they kept in touch with constant messages. Todd was speechless when he was told that one was from the telegraphist Cyrus W. Field, who had laid the first successful cables across the Atlantic. The message said: 'I shake hands with Mr Todd in admiration of his energy and ability on piercing the great continent of Australia uniting the far east with the far west by the electric code of commercial enterprise and loving unity.'

Another was from Todd's brother Henry, now based at the Greenwich Observatory, whom he hadn't seen for nearly twenty years. Henry quoted a poem written in tribute to Samuel Morse and tapped it out himself:

> But one morning he made him a slender wire,
> As an artist's vision took life and form,
> While he drew from heaven the strange, fierce fire
> That reddens the edge of the midnight storm;
> And he carried it over the mountain's crest,
> And dropped it into the Ocean's breast;
> And Science proclaimed from shore to shore,
> That Time and Space ruled man no more.

He signed it, 'Your loving brother'.

The *Telegraph* newspaper in London wrote, 'Time itself is telegraphed out of existence.'

The menu for the 'the banquet to Charles Todd, officers and men' in Adelaide included boned turkey, roast turkey, roast duck, roast fowl, roast goslings, saddles of mutton, guinea fowl, pea fowl, roast beef, sucking pigs, ox tongues, York hams, pigeon in aspic, 'myoncase' of chicken, lobster salads, veal and ham pies and fillets of veal. For pudding there were Dantzic jellies, Madeira jellies, Maraschino jellies, apricot creams, raspberry creams, cream of Venice, blancmange, tipsy cakes, trifles, fancy pastry, pyramids of pastry, maids of honour, Genoese pastry, fruit tarts, orange jellies, champagne jellies, ices and cherries, strawberries and oranges.

Almost three hundred men squeezed in to feast after surviving for two years on damp, weevil-ridden flour and tinned stew. Three hundred guests met them. The post office had sent over the telegraph instruments, and the lines through to London had been cleared for the whole night. Todd sent the first message to Lord Monck. Two hours later they received a reply.

At the end of the meal, the Governor got up to speak. 'In welcoming back, thank God, in health and undiminished vigour my friend on my right,' he said, turning to Todd, 'who has been of such signal service to the colony . . . in the presence of the man whose earnest purpose and constant counsels have urged the advantages of this work . . . we are here to reward those who have performed.' He then announced that Henry Ayers, the Chief Secretary, had been made a Knight Commander of the Order of St Michael and St George, and would now be Sir Henry, although he had taken office less than six months before. Todd had been made only a Companion of Honour. If he was disappointed, he didn't show it. The leader in the *Advertiser* the next day, commenting on Todd's speech, said: 'Nothing in it was a suggestion of a mean and jealous spirit. He seemed more anxious to speak of the courage, industry and perseverance of the men than of his own achievements.' Queensland, Todd's main rival, had the last word, finally admitting that the new telegraph was: 'One of the astounding facts of modern times which surpass fiction'.

26

·· — — — — ····

The Promised Land

Ed and I had three days to get to Darwin in time for a tea party. The Northern Territory National Trust had decided to hold a fête to celebrate the 125th anniversary of the overland line. We were to be guests of honour and draw the raffle. Then there was the black-tie evening ball with the Northern Territory Institute of Engineers. Civilisation was waiting by the sea, but we didn't have a clean T-shirt between us.

Pine Creek was the next stop on our track, fifty miles from Katherine. It is the only original mining town remaining from the first gold rush, when Todd's telegraph men, bored with waiting for Patterson to arrive, went rooting round for opals and found traces of gold in the gullies. As they began to dig the first pole shafts, a nugget was found. From then on, it was impossible to prevent most of the men spending their spare time panning.

As news filtered down the line, gold diggers from New South Wales and Queensland starting making their way north, using the newly constructed poles to guide them. They were met by Chinese from the Kwantung province making their way south from Darwin, pushing their wheelbarrows filled with picks, shovels and cast-iron pans. After years of watching Melbourne grow plump on the yellow metal, and Queensland become hysterical when gold was discovered in the white quartz reefs at Gympie, South Australia had finally got its own gold frenzy in the Northern Territory.

Thus the telegraph's first news from this exotic new land was to sing seductively of gold. A few old telegraph hands formed a company when they returned to Adelaide, and sent a party of prospectors to see just how much gold lay in the hills. The first boat lost her mast and rudder but managed to limp into Darwin harbour. The men then floated up the river on a raft of casks and, with ten horses, set off into the jungle. After one hundred miles, they'd given up hope. On a stifling afternoon, one of the prospectors threw his pick into the ground and cursed all gold. And then he saw it, glinting where his pickaxe had fallen. They sent the news down the telegraph the next day. They had found a lode two hundred feet deep, eleven feet wide and twenty miles long.

Even Todd, who had always promised that his line would open up new lands, was astounded by the speed with which Union Reefs, Burrundie, Grove Hill, Fountain Head, Yam Creek, Spring Hill, Howley, Bridge Creek and Zapopan grew up with tram drivers, farriers, hoteliers, doctors and cooks all pegging claims, filling their calico bags with nuggets and doing a little shoeing of horses or pulling of teeth on the side. They bought tarpaulins, moleskins, mesh sieves and shovels in Darwin and headed down river. When rum ran short, they made their own version, using methylated spirit and kerosene mixed with Worcester sauce and sugar, or gin, vinegar and saltpetre.

A pick and shovel ten feet down turned up four thousand ounces at Pine Creek. One digger rode in with his pack-bag bulging with 750 ounces of gold. The line was jammed with mining agents, company promoters and prospectors, sending urgent, tempting messages down the line. Some of Todd's men at the post office in Adelaide couldn't bear sending so many messages about others making fortunes, and left their keys in search of their own. A telegraph man named Howley founded the richest mine in the north. Others banded together to form companies in the Lady Alice Mine and the Telegraph Mine. They retrieved eighty ounces from ten tons of milky quartz at Lady Alice. Even Darwin's Resident had 'gone to the reefs'.

One linesman, Fletcher, made his money by throwing a shackle over the telegraph line and charging for messages home about fortunes won and nuggets just missed. Todd watched as his men cashed in thousands, put their money back in new mines, bickered, sued each other and lost the lot. He had no interest in gold and, like the public servant he was, cared more for status than easy money.

Todd had also remembered the Wet. The miners, like the telegraph men, were soon cursing the weather. At Pine Creek, three inches of rain fell in ten minutes. Diggers, nourished only by rum and greed, began to die of fevers, dysentery, malaria or starvation. Supplies were not getting through. The news on the line was grim. Five people a week were dying in the camps. Todd also heard about the messages being sent to the doctors in Darwin and Adelaide. One message went: 'Man brought here prostrate, skin cold and clammy, pulse feeble, have given brandy and eggs, eyes now glassy and vacant . . . please advise.' Dr Millner in Darwin replied, 'Give eggs and brandy, extract of malt, almost sure to die.'

In the end, the isolation, typhoid, heat and wet conspired against the Pine Creek area. By the turn of the century, most machinery was turning to rust, bullfrogs inhabited the waterlogged shafts and all that remained of Lady Alice were scattered bottles, a few harnesses and tent pegs, a cherry pipe or two and some discarded sheath-knives.

Today, we had been told, bungalows had replaced the tents in Pine Creek, and the passion and greed had been smothered by bourgeois respectability. Determined to see for ourselves, we turned back on to the Stuart Highway, already shiny in the early morning heat. As we pushed north, the camper-van swaying as we explored its top speed on the bitumen, the vegetation began to thicken by the roadside. But despite the map's promises, there was no sign of any exotic rainforest. It was, and Todd would have appreciated the irony, far too dry.

By the time we reached Pine Creek, the mines' mullock heaps were ablaze in the midday sun. A few truckers had drifted into the main street, switched off their engines, and were dozing in their cabins, while dogs lay comatose in the shade by the wheels. Otherwise the place seemed deserted. The water gardens built to lure intrepid tour groups had dried up and the town centre lay like a concrete husk in the heat. The museum, in which we had been told to expect an interactive historical experience, complete with the opportunity to send messages down the fabled telegraph line, was shut and bolted. The Hard Rock Café was closed.

We wandered into Eddie Ah Toy's general store, which looked unlikely to sell a ball-dress for the engineer's hop in Darwin. A notice on the door said, 'Next Doctor's visit 29th September' – a month away. Inside, sausages were rotating under a grill. The Shell service station sold chamber pots, Imperial Camp Pies, Devon Rolls, tinned mushy peas, ice-cream topping, garden watering systems, battered crab-sticks, an electric blanket, a crib, a school satchel, baking powder, car spray, Vick's VapoRub, a showercap, a book called *Eight Million Ways to Die*, and right at the back, men's grey slip-on shoes. I bought a pair for Ed – he couldn't wear trainers to the dance – and carried them back triumphant to the car, along with a black bow tie with pink dots which I'd discovered next to the size 42F salmon-pink bra.

Back on the road, it was a trip through a B-movie library – we passed signs to Rum Jungle, Tortilla Flats, the Lost City and Batchelor. We were heading for Litchfield National Park, just off the old telegraph route, to sit under woollybutt trees and admire the acacias and cycads. This was our reward for making the trek over the endless dry miles of the centre. Ed and I were determined to wallow in tropical forests, and splash under waterfalls. We wanted to find the red-tailed black cockatoo and his sulphur-crested cousin, whose feathers the men had collected to make an evening bag for Mrs Todd with a matching cummerbund for her husband. We hoped for a glimpse of the antilopine wallaroo, a large member of the kangaroo family, whose fur could be turned into hats that would keep off the worst of the Wet. We looked forward to the contrast with the Red Centre. This was the

Garden of Eden after our trip through the lands of Cain.

In the 1870s, the area had seemed so rich with copper and tin, and the flat tablelands so fecund with green speargrass, that several men applied for pastoral leases when the line was completed. They stayed on among the giant magnetic termite mounds, battling to keep the jungle's tendrils at arm's length. Now all the settlers have gone, and the jungle has reclaimed a 650 sq. km. national park. It is not exactly tropical, but after the Centre it was like stumbling on a long-forgotten botanic garden.

'No pets or firearms' said the sign, as we swung off into the dirt; 'Four-Wheel-Drive Only'. Our camper-van lurched from crater to bump until we found a tree to leave it under. Then we set out on foot past termite mounds arrayed like gravestones, all pointing north-south, through the blacksoil plains and under the dappled woollybutts. No map needed, or so we thought. We'd just wander in the direction of Florence Falls. Three hours later, we were still clambering through scrub, tripping over roots, our arms lacerated. Then suddenly Ed saw something bubbling in the waters of the stream we were following. 'It's washing-up liquid,' he said. 'We must be near a camp.' Ten minutes later, we'd found a larger tributary, and then we saw a bread-knife glinting in the sun. A few yards further on, there was a neat sign pointing to the falls a mile away, a cluster of camper-vans and another warning – 'Keep to Track'.

We followed the markers to the falls, which fell noisily into a deep, shady pool. Did crocodiles live in this part of the park? we wondered, remembering that *Crocodile Dundee* was filmed nearby, albeit with plastic reptiles. But this waterhole looked too domestic, clear and calm, with fish swimming right by the bank. Two children were playing under the waterfall, while their mother was sitting on the rocks, blow-drying her hair with a miniature fan, balancing a mirror on her lap and applying lipstick.

In T-shirts and shorts we waded past her and plunged in. 'Excuse me,' said the lady, 'this is a public swimming pool. You're not even wearing caps.' As if our hair would clog up the filter system. 'You're new around here,' she guessed. We admitted we were. 'Well, you should wash your feet in the stream before you go messying up the water.' Lynette had been at the camp two weeks with her children, and living the Flintstone dream was beginning to pall. 'You don't have any conditioner?' she asked. We hadn't even touched our Body Shop supply, so we promised she could have it if she lent us a towel. Then we offered her our last bottle of wine if she'd drive us back to our camper.

The camp-site was alive by the time we returned with both vans. Men were searching for firewood, and children were screeching for loo paper. So we headed back to park among the slumbering termite mounds under the full moon. I fried a steak, while Ed marvelled at how expert I had become at cooking. He speculated on whether it was a skill that could be transferred to London, or whether it worked only in the outback.

The next morning, as we picked the weevils out of the last of our Frosties, we knew it was time to head for Darwin. We'd just go for one dip at Tolmer Falls, where the ponds fell one after another into the rainforest. At six a.m. no one else could be around. So we slipped off our clothes and slid naked down the waterfalls, floating in the whirlpools. We threw ourselves from rocks, swung from trees and washed off the grime from the Roper River. Just as I was about to dive-bomb Ed from the top of a gully, I looked round and realised that we were being watched by a group of middle-aged Spanish tourists, and they all had their cameras out. I leapt into the water, joining Ed, who pointed out helpfully that our clothes were down at the last pool.

The tour guide was explaining that a few remaining hippies gathered here for the full moon. We must be from the commune in Darwin, but we weren't dangerous, she told her group – she speculated that we might be performing some religious ceremony. Perhaps a purifying of the spirit. There seemed to be nothing for it but to continue with our pagan ritual until they had left. We swam round the small pool in which we were now confined in complicated figure of eight patterns, trying to appear at one with nature, and willing the tour bus on to its next destination.

Then a group of American teenagers appeared, clutching swimming costumes. The ceremony, we decided, was over. 'Could I possibly borrow a towel?' asked Ed, summoning up the last of his English reserve. A giggling girl handed him a candy striped version and Ed, in loincloth, tried to saunter slowly towards our clothes. By the time we were back at the car, fully clothed but stripped of dignity, we had agreed to head for Darwin. We were only three hours from the coast, the sea and the end of the road.

Darwin, population 67,900, once known as Palmerston, owes its northern pre-eminence to the telegraph. This outpost at the end of the line survived Japanese attacks on the eve of its one hundredth anniversary only to be devastated by Cyclone Tracy thirty years later, on Christmas Day 1974.

Waves of the hopeful landed at Darwin after the telegraph poles were laid, convinced it was the promised land. But few managed to last more than a decade. Only the telegraph men and government administrators provided any continuity, eking out their time playing dominoes and charades, staging boating races and holding concerts on the Residency piano. Two years after the line was connected, a Wesleyan minister finally arrived and built himself a church but even he soon returned to more civilised climes.

Someone is always discovering Darwin and its territory, its colour and beauty and infinite resources, but the love affair never lasts. The paradise can be a hell. 'If this country is settled,' said Stuart, who had grown old and blind to reach it, 'it will be one of the brightest under the crown.' A century later, the American ambassador to Australia wrote, 'The Northern Territory can be built into an Empire.'

But Banjo Patterson had the measure of it, coming from Sydney at the beginning of the century.

They start drinking just before breakfast and stop just before too. Everything good is going to happen after the wet . . . A wild land full of possibilities, millions of miles of splendidly watered country where the grass is sour, rank and worthless; mines of rich ore that it does not pay to extract; quantities of precious stones that have no value . . . I would give a lot to be back at Port Darwin in that curious luke-warm atmosphere and watch the white-sailed pearling-ships . . . to be once more with the BAT company and the Overland Telegraph . . . while the cyclone hummed and buzzed on the horizon.

When Todd first arrived in Darwin, he wrote of its discovery, 'The stars danced the day they discovered this new port.' From HMS *Beagle* in September 1839, John Lort Stokes wrote, 'Brilliant meteors fell in the evening of the 20th, a long train of light visible for ten seconds.' He called the port 'Darwin', after his mentor on an earlier, more famous trip with the *Beagle*. But when Charles Darwin started denouncing Adam and Eve, the first settlers quickly changed it to Palmerston.

The front gate of Australia, the ear of the nation, never fulfilled her promise to become the richest city on the continent. Only five years after the telegraph was built, the politicians in Adelaide were bribing men to go north with promise of free land and passage from as far afield as Cornwall and Devon. But there were no takers. The Mennonites fleeing from Russia were offered a berth in the new territory but they refused. Hindus from India were given promises of kind treatment. The Japanese needed to prove only that they could bring a year's provisions.

If typhoons didn't shatter the houses, the white ants got them after they'd gnawed down the telegraph poles. They could eat pistols, letters, jails and billiard balls according to the *Northern Territory Times*, the colony's chronicler. By 1879, the statistics told the tale: Births, 2; marriages, 4; deaths, 154. A Chinese man produced a cabbage four feet across, an orange tree had seventy oranges on its boughs. A German botanist made a garden of English lawns, sugar cane, maize and coffee. No fertiliser was needed. Yet the effort required to clear the ground was so immense that most of the population was still eating bully beef. The century's end saw a sugar rush, followed by a coffee crush and a cattle charge. The town was replanned, with streets named after the first telegraph men. Then the cattle tick arrived, imported on the sirloin steak for the BAT officers in Darwin. And it was back to square one.

Ed and I, sitting in the mall, concluded that we were part of a late twentieth-century information wave. Unlike Adelaide, this was a true southern hemisphere melting pot, in which no single race, colour or creed could claim dominance. Norwegians were sipping cappuccinos with Nigerians, Chinese were jogging with Brazilians in a squeaky clean concrete town garlanded with bougainvillea. Modern Darwin has staked its claim as a chill-out zone between Australia and the exotic East. But despite this role as a cultural clearing house, it remains strangely cut off.

Some things don't change at all. On the day we arrived, the headline in the *Sunday Territorian* was 'Rail Link work to start in '98'. Finally, Australia's various federal and local governments had decided to complete the rail line between Darwin and Alice Springs. Or so they said; locals told us that the link had been announced before, and had always come to nothing.

We scoured the mall for a dress, a white shirt and a swimsuit, before depositing our van on the Esplanade outside the most expensive hotel in Darwin. Open-plan, smelling faintly of the sea, with striking modern art on the walls, shrimp salad on the room-service menu and a king-size bed with proper springs, it made us homesick for our mobile home-from-home. Ed, however, ruled out leaving our clothes in the room and sleeping in the van on the ocean front.

Putting on my newly acquired floral dress, we set out for the National Trust's tea party. Iced tea was being served on the lawn of one of Darwin's few pre-war houses, a two-storey, open-plan structure on the seafront, which doubled as the National Trust's headquarters. Cyclone Tracy obliterated most of the telegraph remnants along with the town, the apologetic members of the Trust explained.

There was no air conditioning, as that would be out of keeping with the period, and as a result there was an ugly crush in the tiny kitchenette, the only room permitted its own cooler. Out on the lawn, Ed and I were introduced as Mr and Mrs Todd, and went round shaking hands, like the squire's guests at an English vicarage fête.

Then it was time for the speech. This time I found that Ed had taken a firm grip on my arm, and was propelling me to the front. I tried to recall the rousing speeches that Todd had delivered to inspire his men, but succeeded only in admitting how very pleased I was to have made it safely across Australia, and to have the opportunity to explore this final chapter of my telegraphic heritage. Everyone clapped politely, and Ed handed me some National Trust white wine from a box next to the home-made cucumber sandwiches.

Before slipping away into the night, we arranged to meet one of the committee, David Carment, Professor of History at the Northern Territory University. He had promised to apply a rigorous academic approach to this final chapter on the telegraph. The following morning, as we walked down the front in search of the point at which BAT's underwater cable had been hauled ashore, David admitted that, intellectually, Darwin was as isolated as ever. He used the modern-day telegraph, the Internet, to stay in touch.

We admired the waterfront where the cable came ashore, and where it was later cut to frustrate the Japanese. David's father had been a prisoner in a Japanese prisoner-of-war camp, and had taken many years to recover, as Darwin had healed slowly and incompletely from their bombs. The BAT buildings, with their ballroom, tennis courts, carriage house and billiard room, had made way for the new Parliament building. The telegraph bell, rung when a ship had sunk, was lost. We located a plaque where the first pole was erected.

But then the history professor left the history behind. We toured modern Darwin with him, admiring his scenic bachelor beach condo, exclaiming over the modern buildings of the university and eating bagels in Darwin's Museum of Arts and Sciences. David was a typical Territorian. Open, generous, a little defensive, he had come for a short break and stayed on, a worthy successor to those early telegraph men, tapping at night on his computer as they tapped out their lives down an earlier wire.

After lunch, we drove into the suburbs to see Peter Forrest, a local historian, who lived in a wooden house on stilts near the beach. Forrest could discuss telegraph poles for hours. Repeater stations were his stately homes. 'Did you get the one just outside the Dunmara Roadhouse?' he asked. 'Did you see the remnants of the *Young Australian?*'

Forrest was ambivalent about his chosen home. Facts poured out. Eighty-two cents out of every dollar spent in Darwin comes as grants from the federal government in Canberra. An average resident stays five years. 'Our role is still to allow people in Sydney to sleep safe in their beds secure in the knowledge that the yellow peril can't sneak in unannounced,' he laughed.

For dinner we went to one of the best Thai restaurants anywhere in the world, set amidst Darwin's cheesecloth and sarong backpacker district. Devouring hot spiced swordfish with cashew and lemongrass at an outside table, we had come a long way from that rejected pie floater on a dark street in Adelaide. But even on the edge of the red dunes of the Simpson Desert we had never felt so cut off from the rest of the world as we did on that isolated Darwin sidewalk.

27

· · _ _ _ _ _ _ · · ·

Just Desserts

Our journey was over. Todd thought his was, too. Then an MP called Charles Mann stood up in the House of Assembly and condemned the line as shoddy and insubstantial. Worse, he said it would soon collapse. Todd was indignant. He couldn't understand where the rumours had started.

A few days later Todd was sent a copy of the *Brisbane Telegraph*, dated 8 November 1872, seven days before the banquet. Inside was an article about 'the work carried out by South Australia'. Todd read further to discover:

> From information from Mr Patterson we learn that the makeshift line at present used is so badly built that there is no possibility of relying on it for any lengthy period . . . Although the fact is not officially admitted it is well known in Adelaide that the line as erected is only regarded as temporary, and it is also perfectly understood that, if some portions of the country through which it passes should be flooded, there is no possibility whatever of effecting repairs . . . whatever the paltry jealousy the other colonies may have evinced towards Queensland should not prevent our Government from using every endeavour to secure an

alternative line of communication.

Todd was apoplectic. When he called Patterson into his office, his former colleague denied all knowledge of the report. But Todd had heard rumours that the engineer had been criticising the line to anyone who would listen, and he'd been seen holding a conversation with Charles Mann in Rundle Street.

Todd gathered all his overseers. They were equally appalled at Patterson's allegations, insisting that each part of the line would hold true, except those built entirely with wooden poles, which were now being replaced. Minutes taken of this meeting record that the overseers couldn't understand why 'one person, for purposes which are inexplicable, chose to promulgate reports so untruthful'.

The *Advertiser*, Todd's staunchest supporter, and the government, his less reliable one, were both against Patterson. Ayers, who had only just received his knighthood, made it clear that Patterson would have to recant or he wouldn't receive a bonus. Patterson at first sent a justification, but soon his letters became more grovelling, and only when he finally admitted that Australia now had the most modern telegraph system in the world was he allowed his £500. After painful discussions in the Assembly on the huge extra cost of the line, the officers had received in bonuses between £150 and £50, petty officers £20 and the men between £15 and £10.

Todd's £1,000 was not enough to build the extra rooms once he had paid off his debts. Alice's holiday had to be delayed. But he was given due credit. Sir Henry Ayers told Parliament, 'Mr Todd's desire had been principally for the men to be rewarded . . . He has ignored himself both in giving merit to those who have assisted the work, and in suggesting the rewards, and has shown the most unselfish spirit that any man could possibly feel.' The Hon. W. Morgan was more blunt: 'All agreed that his conduct was most unselfish. He had made no terms for himself and was a great booby for not doing so.'

But Todd, the career civil servant, was more concerned for his reputation than the money, and was amused enough by the ditties composed about him that he sent one to his brother in England, a copy of which was found in one of Todd's old books.

Little Toddlekins spoke to the ministers all,
He spoke from his heart so bold,
Send me, send me, to the North Countree,
I care not for silver or gold.

Of gold and silver I think not now,
For my heart's in that North Countree,
By night, by day I toil alway
To finish the work – quoth he.

The Ministers chuckled – the Ministers smiled,
They thought – he is jolly green
But he's lots of pluck has our Toddlekins mild
And he'll make all things serene.

Away to the North bold Toddlekins went,
Away to that North Countree.
And the wire was hung and the bells were rung
And the people went mad with glee.

The Ministers haggled, the Ministers frowned,
Quoth they he's a C.M.G.
So we'll fob him off with a thousand pounds
For his work in the North Countree.

And a bonus we'll give to the Northern men,
For they'd many a breakdown there.
With store of gold they shall be consoled,
For the rest we do not care.

Though the Central work was in good time done
And the men they toiled full sore,
As they made no fuss 'tis ridiculous
To pay them a penny more.

So down to the House the Ministers went
And the House there with them agreed:
And a paltry sum did they vote to the men
Who toiled in the North Countree.

MORAL OF THE TALE:
Take warning by this Civil Servants all,
When the Government's in distress.
Make a bargain hard for a sure reward,
'Ere you help them out of distress.

If you show your zeal you'll surely feel,
Your labour thrown away,
For they'll hold that success is reward enough,
Without any extra pay.

And beware how you get through your work in time,
Without a breakdown or two.
For unless there's a fuss 'tis ridiculous,
To suppose they'll think of you.

The Treasurer felt no guilt when he moved in the Assembly the adoption of the bonuses. 'The Government thought, all considered, the sum to be a reasonable one. Original estimate had been enormously exceeded,' he said. Todd then asked for the explorer Stuart and his men to be remunerated. This was greeted with derision by most. The Treasurer insisted: 'The Government did not recognise any claims whatever besides those of the persons employed by them in the work of construction. If navigators had not discovered New Holland, we should not have been here, and it has been said that if there had been a tiff between Adam and Eve we should not have existed.' For Stuart it would have been too late. He had died an unknown invalid in 1866 in London, and was buried in Kensal Green Cemetery.

It was only the next day that Mr Mann stood up to criticise the line, saying that he had been informed 'on very good authority indeed' that it would have to be repoled and that 'there was a widespread impression that they had become too jubilant'. Todd not only rallied his overseers, he also sent a terse letter to the Chief Secretary, saying that he was quite happy to countenance an inquiry into the line. Ayers also received a personal note from Todd:

> While I rejoice to be able to exonerate Mr Patterson of all intention to disparage the work and bring it into discredit . . . and can readily understand that anything he may have said in Queensland would be liable to distortion, I cannot help feeling that the reports have to a certain extent prejudiced me in the estimation of the public . . . a result I certainly did not anticipate after intense labour for the past three years.

Todd refused to accept his bonus until the whole matter had been cleared up. The *Advertiser* wrote: 'Throughout the official correspondence Patterson tried desperately to vindicate himself . . . the very obvious course of action, a letter of denial to the Brisbane paper was not taken.'

The Postmaster needed to quell the critics. The government guaranteed the money for 9,000 iron poles to be erected the next year, using the men who were still on the track. Todd also chose two new teams to go to the Roper and personally waved them off from Sydney. He then insisted that 'a small annual sum' was all that was needed to replace gradually the wooden poles in the north.

But two years after the line's completion, Patterson was still griping. He wrote a memorandum on 3 June 1874 to the Chief Secretary attacking

Todd, and insisting that he was the cause of Patterson's sufferings. Todd, in forwarding the memo, merely refuted a few points. The line was still standing, and he thought he had won.

He had forgotten about Queensland. The brasher, bolder colony couldn't bear to rely on South Australia for long. At an inter-colonial conference in Sydney in 1875, the subject came up again. Sir Henry Ayers went as the South's representative. The charges from Adelaide to London of £9 6s. 6d. for twenty words were heavily disputed, so a standard rate of 'one shilling for ten words (exclusive of address and signature), and one penny for each additional word' was agreed. Ayers then read a lengthy report from Todd detailing the strength of the line.

Queensland nevertheless insisted on reviving her initial scheme for a cable from Normanton to Java, claiming that it would be 'foolish' to rely on just one cable. Western Australia advocated its own scheme for a cable to Ceylon. For three years, William Cracknell, the brother of Todd's old assistant Edward Cracknell, plotted and schemed, but at every turn Todd matched him with more iron poles. On 18 May 1875, Todd sent a triumphant note to the Chief Secretary saying that he believed the line would now hold for at least fifty years.

'Owing mainly to the very admirable manner in which the Overland Telegraph has worked and its immunity from interruptions, the negotiations for an alternative cable to Normanton have fallen through,' he wrote jubilantly, 'but, that being the case, it is more than ever necessary to maintain the efficiency of the line to Port Darwin.'

How long did this singing line hold? It was still standing at the turn of the century. Every few days, the service on the landline would be interrupted and once in a while the sea cable would fail, and everyone would complain. In 1876, almost six months were lost due to cable failure. In 1898, a copper wire was strung over the land route, replacing the original galvanised iron. But no one had the energy to build another line. South Australia had a monopoly and guarded it jealously. By 1901, 43,000 miles of telegraph line criss-crossed the country, and six million telegrams were handled every year. In 1902, the final link around the world was made when Vancouver was connected to Australia. The girdle was complete.

But eventually the line had to compete with other forms of communication. The telephone, invented in 1876, was slowly being introduced to the major cities, and Todd spent the 1880s perfecting a local telephone exchange. His telephone number in Adelaide was '3'.

A favourite story in the north is of the night at Powell Creek in 1906 when the operator, Billy Gents, told a group of drovers of the great news. Marconi had perfected radio-telegraphy. Using a long wavelength, he had managed to talk across the Atlantic without a wire. The wireless was born. 'That means you're out of a job, Billy. The writing's on the wall,' they said. 'Not so', Billy replied. 'He'll never get away with it. You can't have the whole world tuning in.'

The whole world did tune in to news and entertainment on the radio, but personal messages still had to go down the line. By the First World War, the kinks had been ironed out, and at Alice Springs the wire had been moved from Temple Bar to the wider Heavitree Gap. In 1907, Sydney and Melbourne were connected by telephone, seven years later Melbourne and Adelaide were linked. In 1930, Australia was finally able to communicate with Britain by telephone. But it was extortionately expensive, and the telegraph was the main route for almost all messages sent abroad. To cope with demand, another copper wire was added, and teleprinters were now used as well as Morse-code operators.

Then, on 19 February 1942, the technicians in Adelaide were carrying out tests on the line to Darwin with the supervisor there, 'Snowy' Halls. After years in the outback repeater stations, Halls had applied to be sent to Darwin for the war. During the fall of Singapore he hadn't left the key for four nights and four days as he relayed messages. On that night, Snowy warned Adelaide: 'The Japs have found us and their bombs are falling like hailstones. I'm getting out of here, see you later.'

The next Adelaide heard was from a divisional engineer in Darwin.

Using an old Morse pocket relay, the kind that Todd had used at Central Mount Stuart, he had run to the outskirts of the town and tapped back the message. Snowy was dead, the station had been hit, the line was down, and they were expecting Japan to invade at any moment. An officer, thinking that the Japanese would want to retain links with the world, waded into the shallows and sabotaged the cable. It would have taken weeks for the Japanese to work out where it had snapped. The Rising Sun's landing craft never hit the beach, but Todd's line was permanently out of action for the first time in seventy years.

After the war the line was fully reconnected, but the telephone was now the great threat. Queensland finally had its way and international telephone calls arrived there, before being directed down the eastern shoreline, hugging the coast as they made their way around to Adelaide. STD calls from Adelaide to Alice Springs went via Sydney, Brisbane and Tennant Creek. But communication remained patchy; as late as 1961 only sixteen international telephone calls were made an hour between Britain and Australia.

In 1975, Telecom, now Telstra, separated from the Post Office. The state monopoly was finally broken in the early 1990s when the private company Optus challenged the publicly owned carrier. The main fibre-optic cable routes now go from Sydney to Auckland, Honolulu and San Francisco, and from Sydney to Guam, Hong Kong and Tokyo.

But the telegraph had led directly to several great communications developments: Bell's telephone, Edison's phonograph and Marconi's wireless, and they were all embraced by an enthusiastic Todd before he died. Telecommunication is defined by the Oxford English Dictionary as 'Communication over long distance, especially by means such as telegraphy, telephony, or broadcasting'. The telegraph's most direct descendant is e-mail, which now allows twenty-first-century Australians to send instantaneous messages around the world. But its most enduring feat was to disentangle communication from transportation.

28

·· —— —— ——··

The Sunburnt Civil Servant

What happened to Charles and Alice? 'Charles Todd began to be mythologised,' according to Kevin Livingston, a historian I met on a rainy afternoon in Ballarat, Victoria. 'Todd was a comparatively young man of forty-three when he embarked on the task that was to bring him fame in his own lifetime, and two years later he was well on the way to becoming a legend thanks to the eulogising of his contemporaries,' said Professor Livingston, who had just completed the antipodean telegraph junkie's bible, *The Wired Nation Continent*. Although the Postmaster-General lived for another thirty-five years, the shadow of his great achievement obscured his later life. Alice was only thirty-six when the line was completed. I had to piece together the rest of my great-great-grandparents' lives from family fragments.

Professor Livingston had been introduced to me by John Jenkin, the man who had saved me when I lost all my luggage. John, a historian, lived in Melbourne and was more interested in my great-grandfather, William Bragg, who had married Nina, by now renamed Gwenny. When I was still in a jet-lagged haze, he had shown me William's courtship letters to Todd's seventeen-year-old daughter. His wife, Consie, had kindly shared her wardrobe as I waited for my suitcase, and supplied me with endless mangoes and hot buttered toast.

Todd obviously considered Gwenny, the beauty of the family, far too young to marry. When the dashing English scientist, the twenty-three-year-old William Bragg, came out to Adelaide to work at the University, he soon became friends with Gwenny, giving her a pair of gloves for her birthday and sending flowers when she had whooping cough. Todd decided to pack his daughter off to Tasmania to keep her out of the way of this young Englishman, with her brother Charlie as chaperone. But Alice, perhaps remembering her own engagement, had other ideas. When Professor Bragg came to dinner and explained he might like to take a holiday in Tasmania, she handed him Gwenny's forgotten blue sash, and explained that he would be doing her a great service if he could deliver it to her daughter.

Charlie soon sent a telegram from Tasmania. 'Professor Bragg wants to be engaged to Gwen.' Todd, never comfortable with emotional dilemmas, cabled back unhelpfully, 'Say everything that is kind to both.' Gwenny, like her mother twenty-five years before, was still a little unsure, writing to William, 'You mustn't be always trying. Love is very hard . . . just leave it all to take care of itself.'

William adored the chaotic Todd family. They nicknamed him, 'The 'Fresser'. He wrote home: 'I met the Todd family for the first time. Such a jolly lot they were. Mrs Todd made the household, of course. I was most impressed by her calm statement that she didn't think she could go to the Government house party because she had not a dress fit for it. Such open and unconventional a confession was a surprise to me.' He soon married Gwenny.

A faded newspaper report fell out of one of my great-grandmother's letters, describing a garden fête at the Governor's house. A decorated parasol parade had been arranged by Mrs Todd and the new Mrs Bragg. 'All dressed in snowy white garments, and carrying parasols of infinite variety, and beauty', the young women of Adelaide paraded to the strains of a string band. Miss Waterhouse won first prize, with a parasol of pale green, with white roses and asparagus fern, trimmed with chiffon. Miss Craig came second with a pale blue lattice-work construction with two lovers' knots of white roses, and Miss Fotheringham came third for a Japanese umbrella with white roses, fern and lilies.

But despite helping to arrange this entertainment, Alice was becoming increasingly frail. Gwenny often despaired of her mother, who kept introducing her husband as Mr Boast, not Mr Bragg, and when her son-in-law took her to his laboratories and X-rayed her hand with his new toy she shouted out in terror at the flashing sparks and the smell of ozone. As for Todd, when I looked him up in the *Australian Encyclopaedia* for 1926, he came between 'Tobacco, Wild – noxious plant' and 'Tommy Rough – see Australian salmon'. But other than recounting his feat with the telegraph, the encyclopaedia offered scant information on his later life. After three months spent crossing Australia, my knowledge of Alice and Charles's last years was as flimsy as Miss Fotheringham's parasol.

Pat became obtuse when talking about the Todds after the overland line had been built. Yes, she had heard some rumours about Todd losing his money, but it certainly wasn't in gold and she wasn't going to elaborate. Returning to Adelaide, I holed up in the library, starting with old copies of the *Advertiser.*

Todd, I discovered, could never stop being the perfect civil servant. He had only become Postmaster-General in 1870 and was determined to get it right. His postmen were given scarlet uniforms, the mail carts were painted red and he spent hours amalgamating the telegraph and postal divisions into one smooth service.

But the northern line was still his greatest love. Todd personally chose all the repeater stationmasters, and regularly helped them to while away lonely evenings by playing chess down the wire and swapping jokes. He could recognise each man merely by the style of his Morse code. As the travel writer Ernestine Hill wrote in her book *The Territory*, published in 1951, 'Until 1930 the Overland Telegraph track was the faint midrib of the fifth continent, a ghost of a road, the width of a wagon, running from south to north. The telegraph stations were the great first cause in colonisation of a round million square miles.' They started as shacks in the wilderness, each with two operators, a linesman, a cook, a box to cover the instruments, and a rifle above the key to cover raids. But soon these stations became 'fortresses against the implacable loneliness, citadels of law and order in a lawless, loveless land'. To the early explorers and cattlemen they appeared as mirages. For Ed and me they had often provided the only petrol for two hundred miles.

Todd would send only his best men to the centre. In return for eight shillings a day and rations, the men were expected every night at ten o'clock to receive and repeat the news from England in Morse code. These overland telegraph offices never closed. The Postmaster-General wanted the stations to become thriving villages. He encouraged the men to give board and tucker to any travellers up the track, provide medicine and information and swap stories. The men struggled against nature to keep the line open, sending on messages of births and deaths, discoveries of gold and ores and, later, wars. White ants ate the remaining wooden poles, Aborigines stole the insulators and the wire rusted. Cyclones blew the line down, trees fell on it, lightning struck it, sandhills washed over it and floods engulfed it. But the binary link with the old world would usually be reopened in a couple of days.

The linesmen were constantly monitoring the line, sleeping in their swags, swimming their horses over creeks and climbing the poles to check the insulators. For the rest of their time they mustered the cattle, tended to small vegetable gardens and bartered with the Aborigines. Their wives and children soon joined them. Todd found that many of these outback officials became addicted to the life, and refused to return to work at the GPO. Their neighbours were the other telegraphists, and when the lines were quiet they would often have a chat with Java or Singapore. Tales were passed around the world, such as the one of the German woman who went inside a telegraph office with a dish full of sauerkraut which she wanted telegraphed to her son, who was a soldier. The woman had heard of the soldiers being ordered to the front by telegraph, so surely their supper could go too.

Even in the 1950s, Waldermar 'Wallaby' Holtzte was still out there, keeping the repeater station intact just in case it was ever needed again. At seventy-nine, he was the officer at Powell's Creek. He'd joined the service in Darwin at fourteen, and was soon sent down the line to Daly Waters. When the aeroplane called there in 1930, he applied for a transfer and was sent to the furthest outpost, Tennant Creek. But then there was a gold rush, and 1,000 people joined his oasis. So he went to Powell's Creek, with the logistics of war following close behind him along the new military road.

Todd hadn't finished with his telegraphic meanderings across the continent. With so much of Australia now wired, he was determined to link the last state, Western Australia, to his Grand Project. The Western Australians had been murmuring about getting a line for years after the overland telegraph was built. Ten years later, the colonial secretary, F. P. Barlee, finally got the money together, and asked South Australia to help construct a line from their town of Port Augusta around the coast to Western Australia's Port Eucla.

F. P. Barlee was impressed by the sunburnt civil servant he met in Adelaide. 'At an early stage of the negotiations the valued support and co-operation of that Grand Old Man of South Australia made its influence felt,' he later wrote.

Charles Todd, Postmaster-General, Superintendent of Telegraphs and Government Astronomer, was indeed one of nature's noblemen. Gifted with an all embracing intellect, he was equally at ease whether engaged in abstruse calculations to determine the movement of the planets, solving an electrical problem, or in diplomatic controversy with his ministers. While in private he was always the genial, kindly gentleman, interspersing his conversation with an occasional pun . . . his persistent advocacy of our cause, and his unobtrusive practical suggestions relative to the equipment of the teams, the type of insulators best suited to the climate, and many other constructional details, were extremely useful throughout negotiations.

Todd, now fifty-eight, loved any excuse to get back to the bush. He urged Parliament to help with the necessary funds for the projected 780-mile line and, in a personal letter to Barlee on 30 January 1884, wrote: 'This is a national obligation which our geographical position compels us to fulfil.' By the beginning of 1885, tenders had been accepted, with the government again taking responsibility for some sections. Todd made his friend Knuckey supervisor of the project. Again, they set out with no firm idea of the land to be covered, The drought had set in, so all provisions were taken round the coast by sea, and the horses needed four times more feed than usual. The men had to contend with dense mallee scrub and forty-degree heat, but they now knew the importance of iron poles. Using 12,474 Oppenheimer poles, Knuckey wired almost a thousand miles in under two years. Todd's original estimate for the line was £65,000, the actual cost was £68,205. He had learnt from his earlier mistakes.

While Australia was being laced with thin metal wire, the affairs of the Observatory had been overlooked. Todd felt increasingly embarrassed at the paucity of his information about the stars of the southern hemisphere. He barely had time to read the papers sent from Britain on the latest discoveries, and observations were 'greatly in arrears', according to one of his reports. But in 1874 he managed to get a new equatorial telescope in time for the transit of Venus. He also employed his first assistant, Alexander Ringwood, who had been a surveyor in the Northern Territory. 'I am anxious to see our Observatory popularised as a school of physical science,' he wrote, 'at which regular courses of lectures should be delivered on practical and physical astronomy, navigation, meteorology, magnetism, electricity, heat, light and optics.'

Then there was the time-ball at Semaphore, on the coast ten miles from Adelaide, which was desperately needed for the ships to set their chronometers. Perhaps in recognition of the challenge that had won him the right to come to Australia, Todd spent hours on the ball. It was dropped by a voltaic current from the observatory at one p.m. The ball was 5 foot in diameter, fell through a space of 13 feet, and was 93 feet above sea level. It only failed to fall twice during Todd's career, and for most of that time the error was no more than two or three tenths of a second, making it the most accurate ball in the world.

If Alice thought she would get his remaining time once the GPO and Observatory were running smoothly, she was wrong. Meteorology became Todd's next passion. He was adamant that the number of meteorological stations round South Australia should be increased and that the repeater stations should all keep records of the local weather. Every morning, when in Adelaide, he would wander through the paddock to the dome which housed the big telescope and enter the small lattice-work pagoda to look at the temperature in the shade. He checked on the thermometers turning in the sun, the underground thermometers, the rain gauges and the evaporation tanks, which had a fence to keep the dogs away; although as Lorna said: 'I never felt the readings from this evaporation could have been very exact, for I used to see birds drinking thirstily on hot days, and I myself used to fish for tadpoles there.'

Young clerks also took weather reports five times a day, recording the barometric readings, soil temperatures, wind directions, measurements of solar and terrestrial radiation and the general character of the weather. Each year, Todd would hand in a meteorological report to Parliament often running to more than a hundred pages, detailing storms and drought across the continent from Adelaide to Darwin. Obsessed by his charts and diagrams, he became the Australian father of the uncertain science of predicting the weather.

He was also adamant that Australia should simplify her idiosyncratic local times. Perhaps remembering his early work organising the free distribution of Greenwich Mean Time across Britain, he spent ten years arguing for standard time zones, which were finally introduced in 1895.

To the west of the garden, Todd bought another paddock, which he filled with horses, cows, and a duck pond. Everyone was surprised when Alice became pregnant again. Lorna was born in 1877, and the red-headed child soon became the family favourite. Coming eight years after her other brothers and sisters, she had only a nursery housemaid, 'who seemed too busy housemaiding to do much for me', so she was allowed to run wild in the Observatory. Lorna wrote, 'In winter, the marsh mallows grew taller than I was and made excellent cover on Saturday mornings from which I could watch my young nurse searching for me to wash my hair.'

The children often slept in the garden, despite the protestations of Alice's mother. 'Many a night we camped on the lawn outside the dome – I still hear the tramp made by the feet of the parties who came to look through the telescope and to see, as my father put it, the glories of the heavens. As we lay on the cool grass we would gaze up at the stars and my father would come down from the dome and explain the wonders to us.'

Alice stopped asking for her dressing room, although she still hoped idly for a holiday if her husband would ever stop working. Todd refused to discuss a transfer back to England; he had made his name in Adelaide and didn't want to return to the dank, upright mother country. But he did rent a small cottage for the family at Port Elliot, where the children would go for three months at a time to escape the Adelaide summer heat. Alice remained frustrated by his endless new projects, but she enjoyed their new status, and in her forties began to accept that she would never return to Cambridge.

'I remember the Observatory as a very happy home,' Lorna writes. 'But it wasn't convenient.' There was no water laid on inside the house, just a large underground tank below the yard with a pump off the veranda, so that in winter they stood in the rain to pump and in summer in the sun. The bathroom was on the back veranda – an enclosure of thin wood housing a huge slate bath 'where my mother and father bathed in cold water all their lives,' according to Lorna. 'It must have been very chilly, as the windows must always be left open because two stonemartins had built a lovely nest of little mud pellets on top of the shower.'

Alice once suggested moving because the house was 'at the poor end of town', but Todd dismissed the idea as an extravagance. Instead, their smart friends made the pilgrimage to them. At one stage, Alice had three hundred calling cards. Lorna remembered her mother's days at home in the 1880s. Carriages would come through the wide open front gates, sweep around the oval flower beds, and deposit grandly dressed matrons with card cases at the front door. 'A ring at the well polished bell would bring a parlour maid in cap and streamers with a silver card tray, to usher them in, while the carriage drove out to wait on West Terrace,' wrote Lorna. 'Here the barefooted, bedraggled boys of the district would congregate, voicing their sharp criticisms of coachmen and footmen in livery.' The Todd children would be expected to entertain the younger guests with archery, croquet and lawn tennis. Lorna recalled;

> Formal dinner parties were the order of the day . . . I loved to creep down the bend of the stairs and watch the procession into the dining room – my father leading with perhaps Mrs Kenion, wife of the Bishop on his arm . . . many a titbit was handed down to me by the maids through the banisters, always ending with a creamy meringue or a plateful of jelly which, when the ladies retired at the coming of the port wine, I would carry away to my bed. My father always described these evenings as 'most enjoyable'. After coffee, cards would be produced for solemn games of whist.

Explorers often found their way to the Observatory, along with botanists and scientists, and Alice would try to slip in her church friends. 'My parents went often to dinner at Government House,' wrote Lorna, 'and because of my father's interesting and amusing personality each Governor of the State in turn became a friend . . . What different days, what placidity and security of social life,' Lorna wrote. The Todds had finally arrived.

29

. . ━ ━ ━   ━ ━ ━ ━ .

Lady Alice

Alice must have been relieved that her husband's gambling days were over, and they could at least enjoy the solid middle-class perquisites of her own childhood. By the late 1870s, the Todds employed Mrs Pickup, the house-keeper; Winnie, the nurse; a maid; Joe, the coachman; Mr Chaston, the Observatory porter, and Mr Willow, the gardener. The boys had graduated from St Peter's and the eldest girls were finishing Miss Senna's Seminary for Young Ladies. Winnie looked after them all and Joe still drove them to school. Adelaide had become a substantial town, Australia was in constant communication with the rest of the world and the early pioneering days seemed a distant memory. Alice and Charles should have enjoyed an affluent old age, as he supplemented his salary with the income from his investments.

But Todd was living on borrowed money, which may partly account for his reluctance to leave his beloved Adelaide, although he was offered positions in both Canada and Greenwich. Of course he had fallen in love with the bush, the winds and the whole of this strange southern hemi-sphere. He loved the range of his job and his social position. But he also owed a substantial amount of money in Adelaide that he was still paying off when Lorna turned twenty. One historian was convinced that Todd had speculated on gold, but I could find no evidence that he had blown the proceeds of a lifetime of hard work in pursuit of the yellow metal. Nor does he appear to have lost it at the card tables. Todd liked a game of rummy and a few drinks, but he never played for money.

Then I received a call from Frank and Musgrove Warwick. They'd heard me on the radio and had pinpointed me on their lovingly maintained family tree. Our mutual cousin, Pat, said she had met us. Would we like to come to tea? I tried to work out the connection. They must have been descended from Alice's niece, Fanny, who had married Mr Davies.

Frank's explanation was very straightforward: 'It was their properties that did it, they lost all their money in land.' Todd and Davies Senior's two stations, Mattawilungula and Moonarcoe, had been a financial flop. Neither man had any time to visit them and the managers kept asking for more money for feed. As Todd became more expert at his meteorological observations, he must have begun to realise that they'd been the worst bet of his life. By the early 1880s, the region's lakes had turned to salt, and the properties were worthless.

I found an article by Lorna which seemed to confirm the prognosis. 'Land in the city in those days could have been bought for a mere song, but my father mistakenly went in for station property, thus losing by degrees both his and my mother's money.' Todd would have seen the green pastures when building lines around Gawler. In those two freak years, when it rained more than it had for a generation or would again in the lifetime of the men who built the telegraph, the desert could resemble a Sussex meadow. The Postmaster-General must have thought that, having travelled the country, he should pick out some properties in which to sink his £1,000 bonus, as well as his wife's and his mother-in-law's small inheritance. Lorna explained: 'In the terrible droughts of the early 1870s and 1880s, so many of the banks failed and station owners suffered heavily. We were not out of debt until I was 20.' In another article, she adds,

What a to-do! No civil servant was allowed to go insolvent. My father

had no security to offer but his word. The Observatory home was of
course a Government building. A friend, Henry Scott, lent him
thousands of pounds, and the only security was his word of honour. In
spite of this financial disaster I never remember money being spoken of
in our family. Perhaps because there was none to speak of.

The parties Lorna speaks of so warmly were actually few and far
between by the 1880s. Knowing his father to be in increasing financial
difficulties, Charles, the eldest son, broke off his training to become a
surgeon in England and went instead to Berlin, where he knew he could
get a qualification more rapidly. He hoped then to be able to return to
Adelaide to help the family. Three years later, he came back with his less
illustrious degree, put up a plaque in the cottage opposite the Observatory
and started practising as a GP. Hedley left school at sixteen and went to
the stock and station agents Elder, Smith & Co. Lorna wrote, 'He was very
good at finance and it was mainly due to him that the family pulled
through.' She remembered her parents becoming 'very old and occupied'.
Todd couldn't bear wasting his time over the detail of his domestic
finances, and Alice sat in the schoolroom endlessly darning stockings, and
knitting white shawls which were so fine she could slip them through her
wedding ring.

Alice's mother's money had provided the family lifeline, but she died in
1875. A frugal woman, Mrs Bell had known how to cling on to middle-class
respectability despite their reduced circumstances. She was a Dissenter of
the 'narrowest, fiercest kind', according to Lorna, who wore handed-down
black mourning sashes, even though she wasn't born until two years after
her grandmother's death.

Mrs Anthony Hall kept soldiering on, still complaining about her
rheumatism after twenty years, and scolding Alice for her scatty ways. When
Lorna was born, my great-grandmother Gwendoline still hadn't been
christened. Mrs Hall insisted on the name Gillam for the youngest child, but
Alice added Lorna from the book *Lorna Doone*, a gesture perhaps of

defiance, because her mother and Mrs Hall had so disapproved of her reading novels.

But Alice could be an attentive and adoring mother, especially when her children were ill. 'I remember her most nursing Gwen and myself through scarlet fever. There were no nurses, of course, and few precautions were taken. A sheet soaked in strong carbolic was hung at our door,' Lorna wrote.

My mother slept on the balcony on a hard sofa outside the windows. It was summer and the heat that year was terrific; long hot days and longer hot nights, but because of the infection my mother never left the balcony or our bedroom. For six weeks doctor Goss kept us on beef tea and milk and soda. On the first day we were to be allowed solid food, the joint of beef went bad. It was the only time during that long hot six weeks I saw how tired and worried my mother looked.

Having lost so many brothers and sisters, Alice must have been constantly concerned that something would happen to one of her children.

Marrying off the daughters became a problem. They had no dowries. So in 1885, when Todd went to the International Telegraph Conference in Berlin, he took with him not Alice, but Lizzie, in the hope that she might find a suitable husband in England. He introduced her to Mr Oppenheimer, of the Oppenheimer pole family, but Lizzie thought him too old. While staying with her cousins in Cambridge, she met a solicitor called Charles Squires, and, after a visit back to Adelaide, returned to England to be married. Alice sent her parlourmaid back with her in an echo of her mother's generous gift. Lizzie never regretted not marrying for money, writing to one daughter, 'I've always been glad, dear, that I haven't been given too much, I might not have been grateful enough.'

It was decided that Lorna need not go to school. 'Economy was essential and my distinct distaste for learning anything discouraged my mother.' Alice did her best to keep to a tight budget. She never bought herself a lamb chop, insisting that she was happy to eat the fat off everyone else's, and she always took the most gristly part of the Sunday joint. Once a month, she would go to the GPO, often with Lorna, to collect her small housekeeping cheque from her husband. She would invariably overspend before the end of the month, and have to return for a top-up, to Todd's frustration.

Lorna remembered one scorching Christmas Eve. Her mother always felt most melancholic at Christmas. The decorated gum trees, glass icicles and even the occasional pine branch on the verandas reminded her of home and skating on the Fens. She was calling for 'a little extra' at the post office for presents. As they walked down Franklin Street, past the squat red-brick cottages and plots of abandoned land, her mother noticed that some of the cottages which normally sold tapes, needles and cottons had special Christmas displays, with a pine branch dabbed in cotton wool, a stucco Father Christmas, tiny Japanese paper umbrellas, and some small wooden toys scattered on dyed green wood shavings. 'My mother was thrilled by the sight. "So much trouble and ingenuity," she said. "Just like the Cambridge marketplace. We must go in and buy a little something to show our appreciation."'

Alice's shopping-day routine was always the same. With her cheque from the GPO she would go to Finlayson the grocer, while Lorna sat on a high stool and was fed chocolates. Then she would walk past the groves of gum trees to Hay & Hans. 'I can see that rather dark shop now – two long cedar counters running the entire length, behind which stood elderly men assistants in black alpaca coats, white shirt fronts and black ties,' wrote Lorna. Alice liked to wander past the milliner's counter, where 'all the latest models from home' were displayed. She would look longingly at these caps and bonnets, but she contented herself with ordering a pattern to be sent home for her daughters, and the occasional trimming.

From there they would pass into the Chinese district, where Alice would buy lanterns and Lorna would look out for the Italian pedalling his tricycle laden with pink and white sweets. On the way home, Alice would have to hurry Lorna past the raucous women dressed in black bonnets with garish feathers selling beer from large white jugs. After that there would be the 'larrakins' to contend with, as they neared home and the poorer district. These boys in black bowlers would demand money for their concertina-playing and Alice always felt obliged to slip them something.

She also turned a blind eye to the fruit in the orchard being raided. Drunks who turned up at the door would be welcomed in and given strong, sweet black coffee. Another couple of waifs she adopted were the Cokers. Mr Coker had lost a leg in the Crimean War and would practise his marching every day in his yard. Mrs Coker would come to talk about the old country and complain about her one-room cottage. Then one day Mrs Coker ran in admitting she had tried to stab her husband to death. He had dug up the only rose in the yard to give himself more marching room. Alice invited her to stay, along with the English dressmaker, Miss Parsons. 'Whenever she had nowhere to go Miss Parsons came to the Observatory,' wrote Lorna, 'sometimes for a month at a time, arriving in time for breakfast, an unforgettable little figure in a skirt so wide it looked as if it must have been made for a crinoline.'

Alice would have to find her something to make. 'If there was calico for underclothes, there was no trimming. I can see my mother now with a flushed face turning out the old cretonne-covered ottoman in the school-room in the frantic endeavour to find something for Mrs Parsons to begin on.' Perhaps because she still felt so homesick, Alice was always trying to help those more vulnerable. 'Marry for love,' she would tell her children. 'Everything else is so fickle.'

In the summer, she would invite 'anyone who couldn't afford a seaside holiday' to come to Port Elliot. Lorna adored those holidays. 'Our greatest summer treat was to hire a spring dray and go for a picnic drive to the Bluff. Once I remember our being stuck in loose sand and the dray nearly turning over. A basket full of cutlery fell out and we spent much time in digging for the slender Georgian silver spoons which my mother had brought out from England.' Alice had gone down on her hands and knees but couldn't find them all. 'Don't mind the spoons, dears,' she finally said. 'I so seldom have time for a jaunt.' Lorna, recalling the picnic forty years later, wrote,

'Do you remember the picture in Punch magazine? The young wife sewing a button on her husband's shirt, who, in reply to his remark that their little jaunt to Paris had not come off, said, "Nothing ever does come off, but buttons."'

Alice accommodated her children's friends, colleagues and admirers at Sunday dinner, although she was always worried about having enough food and was perplexed by the fiery conversation. Charles Junior would lecture the family on germs, insisting his mother boil all the milk, Hedley was obsessed by Darwin's theories, and 'Professor Boast' by his X-rays. The talk often turned to Australia, with the young men arguing that the sherry, claret, or white wine that they had just consumed was far inferior to European vineyards. Todd, angry that these boys could slight their new homeland, would invariably leave the table before the port. Alice would give Lorna a glass to take into her father's office to soothe him.

As Todd worked harder and harder to pay off his debts, Alice again turned to religion. She was hurt when Todd insisted that all their daughters were confirmed in the Church of England, but as long as her husband attended church once a week she accepted his decision. She also liked Archdeacon Farr, who gave the C. of E. services at St Luke's. 'The two daughters were the same age as my sisters,' wrote Lorna. 'They worked very hard in the very poor parish. Julia Farr was heard to say she hoped there would be tennis in heaven; she had had so little here. So my mother was always trying to plan tennis parties for her.' Once Lorna remarked how lucky it was that the Congregational chapel was too far out of Port Elliot to walk to. 'It was the only time I remember my mother being really angry with me. She said it had been a great deprivation to her not to be able to attend her own church.' Lorna adds: 'All the same, I knew in my heart she loved the services of the Church of England. She played so beautifully and had what is known as perfect ear. Her Church of England hymn book was much shabbier than the Congregational hymnal.'

Alice was becoming more and more absent-minded. When Todd refused to retire at sixty, she took to playing the piano for hours in his absence. The world began to recognise her husband. In 1886, Cambridge conferred upon him the honorary degree of Master of Arts, and in 1889 he became a Fellow of the Royal Society. He was a member of the Astrological Society and the Institute of Surveyors, on the council of the university and vice-president of the board of trustees of the library and museum. His salary was now £1,000 a year. In 1893, his perseverance paid off. One night he said he had dreamt that Lord Kintore, the Governor of South Australia, had come to his office and said to him: 'Well, old chap, I hope you won't throw it in my face if I offer it to you rather late in the day?' At breakfast, Todd didn't know what it was that the Governor had been going to give him, so the family went back to their eggs. But when he went into the office that morning Lord Kintore was announced. The Governor had come to tell Todd that he had been offered a knighthood. 'Your father should have had it long ago,' Alice told her children. 'I hope he will be pleased but I do not want to be called "My Lady" now, it sounds pretentious.'

Todd may have hoped that this long-awaited accolade might revive his wife and go some way to make up for their straitened circumstances and his absences. She soon received numerous invitations to open bazaars, and at each she was asked to give a speech. Lorna remembers that her mother's words were usually inaudible: 'No one else could hear a word she said but they seemed pleased and happy and the smallest Sunday school pupil would give her a bunch of rather tired flowers.' Alice would then trail round the bazaar spending all the money she had brought with her. Lorna had to hide the fare home in her gloves or her mother would exchange it for yet another pot of jam.

This social activity didn't rekindle Alice's zest. In 1898, when Lorna was twenty-one, and Alice was sixty-two, Alice suffered 'a breakdown and a tired heart'. She became a permanent invalid, and her increasingly frail physical state was matched by a fragile mental condition. She made one last public appearance, for Lorna's coming-out ball at Government House. Descending the stairs to supper on the arm of the Governor of South Australia, Sir F. Buxton, she manage to smile wanly. Dressed in a grey evening dress with a soft pink front, she watched her youngest daughter whirl the night away in a white lace coming-out dress, surrounded by the cream of Adelaide society.

The next day she was ill again, and this time she lapsed into a coma. For a month she lingered on, unconscious. Todd still couldn't bring himself to take a day off work. The installation of the telephone switchboards in the office was reaching a critical stage. A message was sent to him at the GPO when his wife finally died. When Todd returned, Lorna recalled, 'I looked at my father, so broken and rudderless, and realised more than ever the part my mother had played both in his public and private life. She was the one who kept encouraging him when everyone else gave up. She believed in him.'

Alice, however shakily, had kept the family going through Todd's frequent absences, and laughed at his puns when he returned. Her favourite motto was, 'As thy day, so thy strength be'. Lorna wrote, 'How often she must have said this to herself as she battled alone with her life during my father's frequent journeys away from home.' Another time she wrote, 'My mother was always so obliging. But there was a strong determination in her character that didn't appear on the surface.'

The Chief Justice, Premier and the Treasurer all came to the funeral. Todd was back at the post office the next day, but the Observatory was never the same. 'My mother was the quiet centre of our lives. One had only to go into the drawing room in the morning, before the maids had tidied the room, to find all the other chairs facing the one in which she always sat.' Elsewhere, Lorna wrote: 'In just the three years following their arrival in Adelaide my mother had weathered two long separations from my father, and had two babies, ants, mosquitoes, flies and servant bothers to cope with. Surely, she too, all uncomplainingly, had helped in the construction of the lines, but with no excitement and no praise.'

30

...−−− −−−−

A Man Like Todd

I met a man like Todd in the Caribbean. An American scientist, he had gone to the tiny island of Nevis with the Peace Corps to teach the islanders how to keep bees, and had never left. In his house, he proudly told us, were 5,000 books and he even subscribed to the *British Goat Journal*. For twenty years, he had been a linchpin of the island, teaching the schoolchildren simple biology, documenting the rainforest, putting up research teams in his home and running tours into the tropical forest. He showed us how to find wild honey and took us to drink with the locals in the rum shops. Even his physique was the same as Todd's – narrow shoulders, no hips, small regular teeth and round glasses. He laughed at his own puns, loved gadgets and prided himself on his good health, pointing to where his fingertip, like Todd's toe, had been cut off in a childhood accident and explaining how it had grown back. He even had a Patterson whom he loathed, in the shape of another tour guide. Along with the sense that I had found a direct spiritual descendant of my ancestor, came a growing irritation. He was confident he knew everything, from the age of the tree under which we were standing to the key to world peace. He had no intellectual match in this small community and, unchallenged, had grown over-confident of his own opinions. Such certainty had helped Todd to span a continent. It may not always have made him an easy companion. I wondered about this man's wife. 'Oh, she loves the place,' he said. 'She was born here, though she does sometimes wish I'd slow down.'

Did Todd ever worry that he should have brought his wife out to such a hard country? Did he ever regret that his time with Alice always seemed to be snatched?

When Alice died, Todd became an old man overnight. Lorna felt obliged to stay at home as his companion. 'We never again sat in the drawing room at night, but in the dining room with our two cats,' she wrote. As Todd immersed himself further in his work, the suffragette movement was just beginning and Todd embraced the new idea. He soon realised that the Post Office provided ideal employment for women. 'In those Victorian days there were so few careers for gentlewomen that the numbers of downtrodden governesses and dispirited companions depressed him. When the telephone exchange was opened he saw in that the chance for positions for daughters of his friends who were forced to earn their living,' Lorna wrote. 'He also placed widows of his friends and former employees as post mistresses in the small country post offices.'

Lorna remembers her father visiting each country post office once a year using the Post Office express. Todd would plan a tour meticulously. He would often be gone for several days, taking Lorna for companionship. She wrote;

It was really almost a royal progress, we were made so welcome everywhere. We always meant to lunch at the local hotel, but no, whenever we arrived about 12 o'clock, we were pressed to stay for midday dinner. My father was most surprised at the lavishness of the meal. Like many clever men he was very simple-minded in some ways and never suspected what I early found to be the fact, that wires were sent ahead of us. 'The old man is on the warpath,' and fatted calves would be prepared.

Lorna said her father always expected people to live up to his expectations, so they did. 'The work was no doubt a strain on these quite untrained women, especially the balance of accounts. They have often told me of sitting up till the early hours of the morning trying to disentangle Post Office and savings bank moneys. That was the way my father worked and because he never expected failure, gratitude and affection for him ensured success.'

In return for their work, Todd helped their children find jobs in the Post Office. 'I remember well his distress when, shortly after having placed a youth at one of the Northern post offices, a £5 note went missing,' Lorna wrote. 'The postmaster was a trusted officer who had been in charge for years. As he and the lad were the only two at this faraway office, naturally suspicion fell on the latter. My father could not bring himself to recall the lad at once and kept putting off this unpleasant business.' Others at the GPO advised him to sack the boy. But Todd insisted on waiting. On entering his office the next day, he ordered a wire to be sent to the post office asking the Postmaster to pull out the third drawer on the left-hand side of the desk. In a short time the reply flashed back that the £5 note had been found stuck below the drawer. Lorna was adamant that it was Todd's second sight.

Whether second sight or not, these wheezes seemed to keep his men in awe. Up until his seventies Todd was still working full-time. Every night at nine p.m. the telegrams would start arriving from all over the state, reporting the climatic conditions. At ten p.m. the two newspapers, the *Advertiser* and the *Register*, would ring for the weather forecast for the next day. Another project was campaigning for street lighting, and with his son-in-law William Bragg he set up the first Marconi wireless poles in South Australia and sent a message over five miles in 1900.

Todd was still working at the turn of the century and, in 1901, he witnessed Australia become a federation. The federal politicians didn't know what to do. Todd's Post Office department was the only one to run at a profit, and the old man was the only head of department who wasn't a minister. No one had the heart to ask him to retire. They merely pointed out that there could only be one Postmaster-General for Australia, so he was allowed to call himself Deputy-Postmaster-General. The South Australian parliament wanted civil servants to retire at seventy, but, as Todd was still going strong, they thought it tactful to drop the bill as long as Todd wished to remain in office. 'The worst that could be said of him was that it was beneath his dignity to make puns,' said the *Advertiser*, who dubbed the PMG, 'Punmaster General'.

At the age of seventy-eight, when he needed two sticks to walk up the GPO stairs, Todd finally stepped down. It was 1905. This benevolent autocrat had spent sixty-three years as a servant of the Crown. There were now 299 telegraph stations in Australia. His children helped him to pack up the observatory and move to a small home in East Terrace. But Todd would still haunt the GPO and lead small astronomical discussion groups or search out his friends, Sir Samuel Davenport, Joseph Fisher and Sir Langdon Bonython, to discuss politics. 'I suppose you could call me an Observe-a-Tory,' he loved to joke. He was so proud of Adelaide's developments that although he was almost bed-ridden by the time the tram arrived, he insisted on Lorna and his nurse taking him on a tram ride to Pulteney Street.

Every year, a young reporter would be sent to ask old man Todd about his achievements. He always welcomed them, and told them tales of the interior. When asked for his proudest moment, he would talk about the day that he sat on Mount Stuart and sent the first messages across Australia. Then one day the story changed. What was his greatest achievement? the reporter asked. 'When I asked my wife Alice to marry me and she accepted, bravely consenting to share my lot in a new and strange land.'

Todd died while taking the sea air at the Esplanade, Seawall, Semaphore, on 29 January 1910, from gangrene in the left leg resulting from a blocked vein. He was buried next to Alice at North Road Cemetery in Adelaide.

Epilogue

The first drop of rain fell as Ed and I left Darwin. The Wet was about to start, but we'd finished our trip and were flying home. Our hand luggage only had room for the set of anniversary stamps, our Akubra hats and one insulator doll. We left the rest of Australia behind.

Two days later I was back at the *Daily Telegraph* in Canary Wharf. As I drove up to the entrance of this East London dreamscape, I noticed that they'd installed a series of giant clocks. Now, the neighbouring bankers would be able to tell the time around the world. There was New York, Paris, Tokyo, Jakarta, Hong Kong, Frankfurt and right at the end, Alice Springs. Alice's one-bank town wouldn't be much use to the financiers, but it meant that I could check out the time on Todd Mall.

Our suntans quickly wore off. The Akubras looked strange hanging in the bedroom. Occasionally Australian friends would come and stay on the sofa, and we'd reminisce about Violet Crumble chocolate bars. I met a woman who worked in the Harrods barbers' shop who was actually called Alice Springs. 'Springs was my surname, my parents thought it was funny,' she said. The wife of photographer Helmut Lang is also called Alice Springs. 'It's my pseudonym,' she says. 'I'm from the place. It's like being called

London.' I began to feel churlish that I hadn't changed my name.

When Ted Hughes, the Poet Laureate, died, I found one of his poems:

> Take telegraph wires, a lonely moor,
> And fit them together. The thing comes alive in your ear.
> Towns whisper to towns over the heather.

But I knew I would never find any insulators on walks through the English countryside.

My brother came back from a trip to India. 'Did you know there's a memorial to a Todd in Delhi?' he said. 'He was a telegraph man and quelled a mutiny, do you think it was your Todd?' It couldn't have been. But I found out that the two men were cousins who had never met. I received an e-mail from a Hedley. 'I'm your fifth cousin,' he wrote. His family had stayed on in Australia. Reviewing a new restaurant, I found the chef from my Adelaide café in the kitchen. But otherwise the only red ochre we saw was on a paint chart.

Then one autumn morning the telephone rang. Ed and I were asleep in bed. 'Hello Alice Thomson are you there?' a man shouted down the line. 'This is Ian Stanley, the President of the Institute of Engineers in Darwin. Do you know what day it is today? It's the 125th anniversary of Todd's banquet. And you're our representative from Britain. Have you got the champagne and jellies ready?'

'Can I ring you back in five minutes?' I asked blearily.

'No, we need you right now. We've got Alice Springs on the line having a barbecue. Say hello Alice.' A chorus of voices shouted, 'Hello Alice.' 'And Sydney's here too. Come in Sydney.' 'Three cheers for the Toddlekins,' they chorused. Ed started to laugh. 'So tell us what happened to your family, Alice, since the days of your great-great-grandparents? Why did you go back to Britain?'

It's a good question, I thought. Why wasn't I still in Australia? I couldn't imagine us in Sydney. The beaches and barbecues weren't what I missed about the Antipodes. But I wanted to return to the space: the oranges and blues, an Aboriginal family laughing by Todd River, a feral camel eating the outback washing and Ed catching barramundi at the end of a dirt track. I could write articles from Alice Springs, but Ed wasn't ready to chuck in everything and chance his luck running a property.

I knew I would always feel more English than Australian. Ed bought me a thirty-year-old blue Volkswagen bus for Christmas. We planned more camping trips. But the strong sun hadn't broken my British reserve. Ed was never going to wear tight shorts and a singlet and I still hadn't mastered thongs.

Gwendoline and William Bragg, my great-grandparents, never wanted to leave Australia. Gwendoline used to win painting competitions for her watercolours and carved wooden bellows. William ran the physics department at Adelaide University, and spent hours conducting wireless experiments with his father-in-law, Todd. He took part in amateur dramatics and was the best lacrosse player in the colony. They had three children: Willie, Bob and Gwendy. The boys rode their ponies on the beach, climbed the Observatory's pepper tree, practised cricket in the paddock and ate their grandmother Alice's custard from stemmed glasses.

They were young Australians. My grandfather, Willie, was just like Todd, a gauche child but a precocious mathematician. Bob, the younger son, was the outgoing, popular one. Their father was becoming well known for his scientific work. He travelled to Europe to discuss his experiments on X-rays with other scientists, and communicated by letter and telegram. My great-grandfather thought of himself as an Australian, but he finally conceded he would have to return to Britain to continue his research.

He accepted an offer from Leeds University. His wife was miserable, hating the dank, dark winters, in the way that Alice had once felt suffocated by the southern hemisphere's heat. My grandfather went to Cambridge university. Father and son were soon breaking new ground, both working on X-rays and crystals.

Then the First World War broke out. My grandfather was sent to the front in France, where he used sound waves to work out the location of enemy guns. His brother, Bob, joined the 58th Brigade of the Royal Field Artillery. In September 1915, the Padre bicycled out to the front to tell my grandfather that Bob had been killed by a shell at Gallipoli. When he saw the bicycling Padre again a fortnight later, my grandfather assumed it was more bad news. But he had come to tell the 25-year-old that he and his father had jointly won the Nobel Prize for Physics. Willie didn't have the strength to celebrate. The Observatory pepper tree seemed far away.

After the war, my grandfather married another Alice. Their youngest child is my mother Patience. She married David, son of another Nobel prize-winning physicist. None of the four children is a scientist and only one of my brothers makes puns.

My mother remembers her father talking of grandfather Todd, who used to tell the young mathematician: 'Be careful what you set your heart on, for you will surely get it.'

Acknowledgements

This book tells the story of my great-great-grandparents, and my own visit to Australia in their footsteps, but it is by no means a definitive biography. The historical facts are correct, but in places I have extrapolated Charles's and Alice's feelings from contemporary sources and have used family and friends' anecdotes.

In Australia, I was shown enormous generosity and hospitality by a large range of people, many of whom are mentioned in the text. In particular, I would like to thank Ian and Diane Blevin, Professor David Carment, Peter Forrest, John Jenkin, Jane Sloane and my relatives in Adelaide and Sydney. They helped me trace both my roots and my route across the continent.

In addition, I would like to thank the South Australian Tourist Board, The Northern Territory Tourist Board and the Northern Territory National Trust for their help. Valmai Hankel and Valerie Sitters from the State Library in Adelaide provided endless cups of tea, and were extremely patient in handling my requests for research material. I also owe a debt of gratitude, which I will not now be able to repay, to the late Patience Fisher for her repertoire of stories.

I drew on many sources including telegraph men's diaries, parliamentary papers, interviews, newspaper cuttings, letters and books on

Australia. I would particularly like to acknowledge Peter Taylor's *An End to Silence*, and Frank Clune's *Overland Telegraph*. Mrs Symes kindly showed me the draft of her late husband's unfinished biography of Todd and Amanda Boyle of the History Trust in Adelaide helped me to track down Todd's possessions. And I would like to thank the State Library, Adelaide, and Telstra for permission to reproduce illustrations from their archives. I am grateful too to Faber and Faber Ltd for permission to reproduce lines from Ted Hughes' poem 'Telegraph Wires' which first appeared in *Wolfwatching*.

I am especially grateful to those who read drafts of the book. In Australia, Professor David Carment, John Jenkin, and Jane Sloane, and in Britain, Mark Stanway and Clare Brennan all read the manuscript and made invaluable suggestions.

My agent, Georgina Capel, has been supportive and generous from the beginning. My thanks also to my publisher, Alison Samuel and my editor, Jenny Uglow, at Chatto & Windus and Bill Thomas, Vice President of Doubleday in New York, for their patience, enthusiasm and inspired advice.

I wrote much of this book on the island of Nevis in the Caribbean, and would like to thank Catherine and Roger Henderson for their generosity in lending their home to me.

I am indebted to Charles Moore, my editor at the *Daily Telegraph*, for his support and to Sarah Sands for her messages. I would also like to thank Peter Stothard, my former editor at *The Times*, for allowing me to take a sabbatical during which I did much of my research for the book and Brian MacArthur who first encouraged me to go to Australia.

My friends have kept me going, and I am particularly grateful for the help of Samantha Cameron, Hattie Ellis, Catherine Fall, Dean Godson, Jane Hardman, Emily Hare, Sharon Saker, Rachel Sylvester, Susan Wilmot and Petronella Wyatt. I would also like to thank William Dalrymple, Matthew D'Ancona, Amanda Foreman, Lucy Juckes and Andrew Roberts for their professional advice.

My cousins, uncles and aunts were extremely generous in allowing me to pillage freely from our common ancestry, and offered help at every stage. Lucy Adrian, Stephen Bragg and Dick Squires were very patient with my requests for information. My brothers, Hugh and Ben, kept me calm, and my sister, Katie, kept me sane, flew to Nevis and made me laugh.

My greatest debt is to my mother for inspiring me with the story of Alice's proposal, and to my father for being so enthusiastic. Without their encouragement this book could never have been started and without Ed's sense of humour and inspiration it would never had been finished. This book is dedicated to Ed, with love and gratitude.

Index

1. A silhouette of the Bell family. Left to right they are: James, John, Charles, Edward (who joined his father in the family business and was Mayor of Cambridge, 1887–89), Alfred, Mrs. Charlotte Bell, Henry, Alice, Elizabeth, Mr. Edward Bell, William, Charlotte and Sarah

2. Charles and Alice Todd
as a young married couple

3. Adelaide at the end of
the nineteenth century

4. The ceremonial planting
of the first pole of the Overland
Telegraph Line, 15 September 1870

5. Erecting the poles,
contemporary engraving

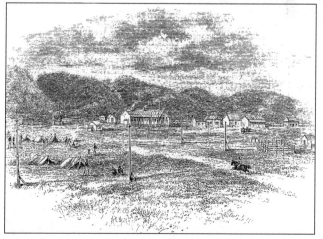

6. The Peake
repeater station
in 1872

7. Early days – while the T-shirts are still white

8. Off the bitumen,
onto the track

9. Ed at the Peake repeater station

10. Ed relaxing two minutes before the Toyota gets bogged down in sand

11. The elusive Simpson's Gap

12. McMinn's sketch of
Alice Springs the day the
station has been put up

13. Alice Springs repeater station in 1872

14. The attack at Barrow Creek, contemporary engraving

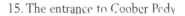
15. The entrance to Coober Pedy

16. The 'angle' or 'angel' pole
– is it real?

17. Camping
with our swags
on the dried out
River Todd

18. The road to Alice Springs

19. Alice outside Alice Springs repeater station

20. The Telegraph fleet off Port Darwin
21. The camp at the Roper River,
contemporary engravings

22. At the Roper River.
From left to right: J.A.G. Little, Robert Patterson, Charles Todd, A.J. Mitchell

23. Bringing the Telegraph ashore in Darwin

24. Ed with a 'feral' pole at Tennant's Creek

25. Making an outback lunch in 110° heat

26. Watching Ed change the tyre after I've crashed the Toyota

27. On the
Roper River

28. The last remnants of
the *Young Australian*

29. We've made it to
Darwin and the Tropics

31. Todd on completion of the line

32. Alice in street dress

33. Charles taking observations at the Observatory

30. Menu for the celebration banquet held at the Town Hall, 15 November 1872

The menu reads:

BANQUET

to Charles Todd, Esq., & Officers & Men

OF THE

Overland Telegraph Construction Party.

TOWN HALL, ADELAIDE.

FRIDAY, NOVEMBER 15, 1872.

Boned Turkeys
Roast Turkeys
Roast Ducks
Roast Fowls
Roast Goslings
Saddles of Mutton
Guinea Fowl
Pea Fowl
Roast Beef
Sucking Pigs
Ox Tongues
York Hams
Pigeon in Aspic
Mayonnaise of Chicken
Lobster Salads
Plain Salads
Veal and Ham Pies
Fillets of Veal

Dantzic Jellies
Madeira Jellies
Maraschino Jellies
Apricot Creams
Raspberry Creams
Cream de Venice
Blanc Mange
Tipsy Cakes
Trifles
Fancy Pastry
Pyramid of Pastry
Maids of Honor
Genoese Pastry
Fruit Tarts
Orange Jellies
Champagne Jellies

Dessert
Strawberries
Oranges
Loquats
Ices, &c., &c.
Cherries

F. W. LINDRUM.

34. Elizabeth, Charlie and Hedley
Todd in the 1860s

35. Lorna Todd as a child

36. The Todd family in 1897, a year before Alice died. From left to right, back row:
Charlie's wife Beatrice, Gwendoline's husband W.H. Bragg, young Alice, Charlie's
sister-in-law Mabel Tower, Hedley, Lorna. Middle row: Alice, Gwendoline, Charlie,
Elizabeth, Hedley's wife Jessie, Charles. Front row: Lawrence Bragg, Frances and
Yolande Tower, Robert Bragg

37. The original portrait of Alice, given to me on my wedding day

38. Sir Charles Todd in his study after Alice's death

39. The plaque commemorating Todd's men